Measurement and Control in Agriculture

Measurement and Control in Agriculture

Sidney W.R. Cox
OBE, BSc, FInstP, FIAgrE

Blackwell
Science

Editorial Offices:
Osney Mead, Oxford OX2 0EL
25 John Street, London WC1N 2BL
23 Ainslie Place, Edinburgh EH3 6AJ
350 Main Street, Malden
MA 02148 5018, USA
54 University Street, Carlton
Victoria 3053, Australia

Other Editorial Offices:

Blackwell Wissenschafts-Verlag GmbH
Kurfürstendamm 57
10707 Berlin, Germany

Blackwell Science KK
MG Kodenmacho Building
7–10 Kodenmacho Nihombashi
Chuo-ku, Tokyo 104, Japan

First published 1997

Set in 10/12.5pt Times
by DP Photosetting, Aylesbury, Bucks
Printed and bound in Great Britain by
the University Press, Cambridge

DISTRIBUTORS

Marston Book Services Ltd
PO Box 269
Abingdon
Oxon OX14 4YN
(*Orders:* Tel: 01235 465500
Fax: 01235 465555)

USA
Blackwell Science, Inc.
Commerce Place
350 Main Street
Malden, MA 02148 5018
(*Orders:* Tel: 800 759 6102
617 388 8250
Fax: 617 388 8255)

Canada
Copp Clark Professional
200 Adelaide Street West, 3rd Floor
Toronto, Ontario M5H 1W7
(*Orders:* Tel: 416 597-1616
800 815 9417
Fax: 416 597 1617)

Australia
Blackwell Science Pty Ltd
54 University Street
Carlton, Victoria 3053
(*Orders:* Tel: 03 9347 0300
Fax: 03 9347 5001)

A catalogue record for this title is available from the British Library

ISBN 0-632-04114-5

Library of Congress
Cataloging-in-Publication Data
is available

Contents

Preface

Almost a decade has passed since my book on farm electronics was published by Blackwell. In that time instrumentation and control systems have been transformed by the non-stop developments of silicon *chip* technology and its associated software. These developments have fundamentally affected all aspects of commerce and industry, including agriculture and horticulture and their support services. The farm office computer is no longer a rare breed; neither is the tractor or combine harvester with built-in electronics for monitoring, control and data acquisition. Electronic systems abound in the dairy sector and in many forms of environmental control. Now crop and livestock producers are beginning to make use of the rapidly developing industrial capabilities of satellite technology and the Internet. Information technology and precision farming have taken root and can be expected to flourish in a computer-literate age, in response to pressures on producers to meet the price, quality and safety standards increasingly expected by customers and demanded by legislators.

In the above setting this book has four objectives. First, by citing examples I have sought to summarise the present state of electronic instrumentation and control in agriculture and horticulture (with an occasional reference to forestry). Second, I have attempted to identify the lines of development that are most likely to influence agriculture and horticulture over the next decade. Third, I have also looked at the historical development of current instrumentation and control techniques. My purpose is to show how ideas and techniques develop, and over what time scales. In my view that perspective is valuable for the insights that it can bring, leading to inspiration for new developments, as many research workers will attest.

My fourth objective is of a different kind, because it concerns fundamental aspects of the measurement process. Although there are now many software packages for rapid automatic analysis of measurement data and tabulation or graphing of the results, it is still necessary to bear in mind the first considerations in any measurement – namely, the accuracy that can be attributed to individual measurements, and the degree to which they are representative of the physical system that is being monitored. These considerations determine the justifiable degree of precision with which output figures can be quoted and compared. I have therefore started the book with an extended chapter on the subject of measurement and measurement sensors generally.

The second chapter is intended to provide a broad, general picture of the development of instrumentation, data transfer and control systems. Thereafter, five chapters cover specific aspects of these technologies to defined areas of agriculture and horticulture.

The book is not written for specialists in instrumentation, control systems or software programming, but for those in colleges, advisory work, manufacture or agricultural and horticultural production who seek an understanding of what electronics instrumentation and computers are capable of in the above areas, both now and in the near future. However, for those interested in studying any of the subject matter in depth I have listed sources of further information at the end of each chapter. Finally, since the instrumentation, electronics and computer world abounds in acronyms I have selected what I think to be the more important ones for the Appendix, although conscious that new candidates will appear year by year.

I wish to record here my debt to the many people and organisations who have supplied me with information and graphic material during the preparation of the book. In many aspects of its production I have received invaluable support from my wife, Monica.

Sidney Cox
April 1997

Chapter 1
The Basics of Measurement

1.1 The measurement process

Standards and accuracy

All measurements derive from primary standards which provide the basis for measurement scales. The means by which the gap between an international primary standard and a measurement at industrial level can be bridged varies from physical quantity to physical quantity. Three examples of commonly measured quantities illustrate this point.

Time

Until 1967 the fundamental unit of time was the day – defined as the average interval between successive noons. However, 300-year records showed that this interval fluctuated by about ± 1 s (second) per year and that over the same period it showed a general decline of about 2 s. Therefore in 1967 the International General Conference on Weights and Measures adopted a new standard – the *atomic* clock. This is based on the frequency with which loosely bound electrons in vaporised caesium switch between two energy states. It is a highly stable 9 192 631 770 Hz. Terrestrial conformity to this standard is achieved by averaging the signals received from atomic clocks in countries around the world, through France (the home of international weights and measures). Satellites for navigation have on-board caesium clocks, providing this so-called *universal* time (a misnomer in view of the relativistic nature of time). Occasional fine adjustments are required to bring it into synchronism with *solar* time, normally at 6-monthly intervals, as anyone with a radio-controlled wall clock will know.

In fact, for most practical purposes the familiar quartz crystal controlled timer is entirely adequate (see Table 1.1) When required, this is easily checked against time information available via the telephone, radio and TV.

[*Note* The SI unit of length (the metre, m) is now defined via time. It is the distance travelled by the light from a helium–neon laser, in a vacuum, during a time interval of 1/299 792 458 s.]

Mass

Here the international standard is the kilogram (kg), which is equal to the mass of

Table 1.1 History of the increasing accuracy in measurement of time.

Year	1700	1750	1800	1850	1900	1925	1945	1960	1975
Time to gain or lose 1 s	3 h	1 day	3 days	10 days	2 wks	1 year	3 years	1000 years	30 000 years
Timekeepers	Pendulum					Quartz crystal		Caesium standards	

the international prototype – a cylinder of platinum-iridium kept at the International Bureau of Weights and Measures (BIPM) in Sèvres, Paris. Numbered copies are held in other countries. These are returned to Sèvres from time to time, to be calibrated against the original. The UK copy is kept at the National Physical Laboratory (NPL), Teddington, where a precision weighing balance is used to compare the Laboratory's reference standards (masses of 1 kg and submultiples of 1 kg) with the NPL copy to about 1 microgram (1 μg). The reference standards are used in their turn to calibrate mass measurement standards held at accredited calibration laboratories, again using precision balances. So, via several stages, any measurement of mass can in principle be made *traceable* back to the international standard.

[*Notes* The terms *mass* and *weight* are often used as if they are interchangeable, because mass is most frequently determined by weighing and indicating scales are graduated in multiples or submultiples of the kilogram. In fact, weight is a *force* measured in newtons (N) in SI units and the weight of a mass M (in kg) is $M \times a$, where a is the acceleration of objects under gravity (9.8 m/s^2 approximately in SI units). The weight of a 1 kg mass is therefore approximately 98 N.

Since this section provides the first use of k for kilo in the book it should be noted that the computer world uses K in the context of kilobytes. However, that K does not represent 1000 but 1024 (see Appendix). To add to the confusion, K also stands for the fundamental unit of temperature (see below).]

Temperature

Although we usually use the Celsius scale for temperature the fundamental unit is the kelvin (K), the unit of thermodynamic temperature, defined as the fraction 1/273.16 of the thermodynamic temperature of the triple point of water. This triple point is the temperature at which a mixture of ice, water and water vapour is in equilibrium – 0.01°C above the *ice point*, 0°C. The precise conversion between °C and K is therefore x°C = (x + 273.15) K.

At the NPL temperatures realised by triple point cells are reproducible to ±0.1 mK. With these cells platinum resistance thermometers and other thermometers can be calibrated over the range 0.5 K to 3000 K, with uncertainties down to a few mK. The platinum resistance thermometer is therefore a useful reference standard for calibration of industrial temperature measuring instruments.

However, where ultimate accuracy is not required a useful approximation to the triple point cell is provided by the simple ice-point apparatus used in schools and colleges – namely a vacuum flask containing a gently stirred mixture of crushed ice and water (distilled). This simple technique also serves well to provide a reference cold-junction temperature in measurements made with thermocouples, although at greater expense the same facility can be provided by electrically-powered reference enclosures without the need for ice and water. A 100°C calibration can also be done with steam generated in another familiar piece of laboratory apparatus, the hypsometer.

In summary, measurement of time normally requires little or no reference back to fundamental standards; verifiable measurement of mass requires traceability to the international kilogram, and temperature measurement can be verified in the same way or, with less accuracy, with simple on-site apparatus.

Accuracy

The three examples above provide an indication of the accuracy of some fundamental standards and substandards. In general, these are orders of magnitude greater than is needed in agriculture and related industries. Nevertheless, consideration of accuracy should never be ignored, if only because measurement to greater accuracy tends to be more costly.

Unfortunately there is a widespread tendency to treat accuracy as an absolute quantity. There are abundant references in the literature to the need for an accurate measurement for a particular purpose or to the development/availability of an accurate method of measurement of some physical property. Unless these comments relate to devices that will count individual items or events, without error, the former specification and the latter claim are both meaningless. The need must be quantified in some way; for example:

- Better than $\pm x\%$ over a given range of the measured physical quantity and under specified conditions (a full specification);
- More accurate than existing methods in a specified application (a minimum target figure).

Equally, a claim for the accuracy of a given measurement or measuring device must be accompanied by a statement of the level of accuracy achievable and the conditions in which that accuracy can be achieved.

This emphasises the importance of dialogue between users/potential users and providers of measurement tools as a means to evaluate the benefits/potential benefits and costs/estimated costs of particular tools with specified accuracy in specified circumstances. That does not guarantee delivery to specification, of course, but it can make the specification more realistic.

Typical accuracies of common measurement methods are given in later sections of this chapter but before leaving this topic the distinction between accuracy and precision needs to be drawn, since these two terms are also sometimes treated as interchangeable. A simple example should suffice to illustrate the distinction: A 500.0 g reference mass is repeatedly placed on a platform weigh scale with a four-digit decimal readout. The scale's readings vary from 499.5 to 500.5. The *precision* of the readings is 0.1 g but the scale's *accuracy* at 500 g is ± 0.5 g, or ± 0.1%.

Statistics

Contrary to a popular view, statistics do not lie, they suffer from their interpretation. As with data processing and computer modelling of systems and processes, the value of the output depends on the quality and extent of the input data.

Statistics are involved in two aspects of measurement:

- Calibration of a method or an instrument, to quantify its accuracy under given conditions;
- Sampling of the physical property being measured, in space, time or both, in order to obtain representative figures for use in management information and decision support systems, or for inputs to automatic control systems.

Calibration statistics

The example of a weigh-scale calibration at one point in its range illustrates the steps required to establish a static calibration of a measuring instrument The graph of a full calibration of a weigh scale might look like Fig. 1.1. In that diagram the magnitude of the interval on the vertical and horizontal scales is the same (0.2 kg). Therefore ideally the calibration curve should be a straight line at 45° to both axes and passing through their origin at 0, as shown. In reality there is usually a loss of linearity with any instrument near its zero output point and imperfections in it may cause the line to be displaced, so that it does not pass through the origin. However, it is not good practice to take measurements at the very low end of an instrument's range, if that can be avoided. A general shift of the calibration line presents a systematic error which can be allowed for or corrected automatically.

The scatter of the measured points in Fig. 1.1 is sufficiently small for a mean line to be drawn through them by eye, minimising the distance between those points and the line over the scale's working range. The situation is not always so simple. The scatter of individual points may be large and the best-fit curve may actually be a curve and not a straight line. In both of these cases statistical methods of curve fitting must be employed. These calculate a best-fit curve from the data, leading on to the definition of pairs of corresponding curves which form the boundaries that contain a required percentage of the measured points. The dotted lines in Fig. 1.1 purport to show the 95% limits calculated in this way, i.e., only 5% of the measured points lie outside their boundaries, on average.

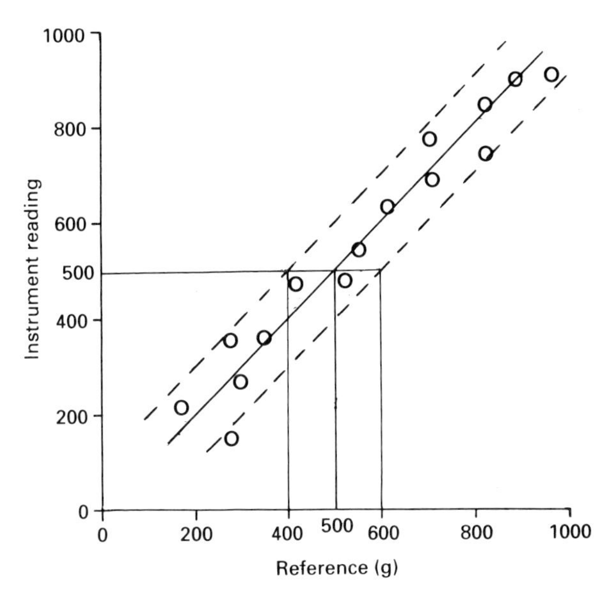

Figure 1.1 Calibration graph: ○, measured points; ——, mean line; - - -, 95% confidence limits.

The statistical techniques involved in these calculations can be found in many student texts. Anyone wishing to acquire an understanding of them should refer to such texts, although there are now many commercial software packages for computer users which automatically process measurement data and generate calibration tables or graphs after performing standard statistical routines.

From the measurement point of view the 95% limits shown in Fig. 1.1 provide a link between the statistical best-fit line or curve and the probable accuracy of an individual measurement. The figure shows that an instrument reading of 500 g could correspond to a mass of anywhere between 400 g and 600 g, as determined by the calibration standard. This corresponds to an error of ±20% at that point in the range. That figure would be higher at lower instrument readings, of course, and it would still be about ± 10% near the upper end of the graph. In addition 5% of the measurements would exceed that error, on average. Clearly, an instrument with that amount of scatter would have a bleak commercial future. Nevertheless, scatter diagrams similar to Fig. 1.1 are sometimes advanced in support of a measurement method – usually by reference to the best-fit curve. It is possible in these cases that individual measurement errors of 20% or more could be tolerated – perhaps because that would be an improvement on existing methods of measurement – but the greater likelihood is that such scatter would invalidate the method as a measurement technique. With most commercial instruments the 95% boundaries would only just be distinguishable from the best-fit line on the scale of Fig. 1.1.

Sampling

Obtaining representative figures on the magnitude of physical quantities is a major problem in almost all areas of agriculture and related industries, whether these quantities are related to outdoor or indoor operations and plants or livestock. They usually vary significantly in both time and space to an extent that can rarely be ignored except at the most general planning level.

Since time scales and the requirements for site-specific detail vary widely from sector to sector and operation to operation these topics are discussed in more detail in later chapters. At this point it is appropriate to comment that standard statistical techniques, including time-series analysis, are available for analysis of variations and trends. Their use is closely related to the accuracy with which operations can be modelled, managed and automatically controlled.

Where measurement of a variable yields a very *noisy* signal there are also standard methods of signal analysis and presentation for extracting information from the data. These include correlation and autocorrelation techniques, to search for regularities in the signal, and power spectrum analysis, to provide a statistical picture of the signal's frequency spectrum. Other mathematics-based analysis deals with the effect of sample length on the quality of the resulting information. These techniques can be studied more fully in many text books on instrumentation and information theory.

1.2 Sensors

General characteristics

Some measurements can be made by direct reference to the magnitude of the physical quantity concerned, using an appropriate scale. Length, area, volume and time are in this category. The magnitudes of many others are measured by analogue means – a simple example is the conversion of temperature into the expansion of the mercury thread in a mercury-in-glass thermometer. The analogue device is termed a sensor and when its output is converted into an electrical signal it is customary to call the combined unit a transducer. The applications of a variety of sensors and transducers are described in later chapters. Here their general characteristics and limitations are outlined.

Sensor dynamics

The first important characteristic of a sensor is the static calibration against the physical quantity that it is required to monitor, as outlined in the preceding section. The way in which its output follows changes in that quantity is another important consideration.

Broadly, sensors respond to such changes in one of two ways, i.e., they behave as a system with a time lag, like that of a mercury-in-glass thermometer (which is a function of its mass and specific heat) or they have the characteristics of a damped mass/spring system. Mathematical treatments of these two systems can be found in many standard engineering textbooks, since they play a part in many dynamic systems, including vehicle and seat suspensions. In order to stay within the confines of this book their basic responses are summarised here in Figs. 1.2 and 1.3(a)-(c).

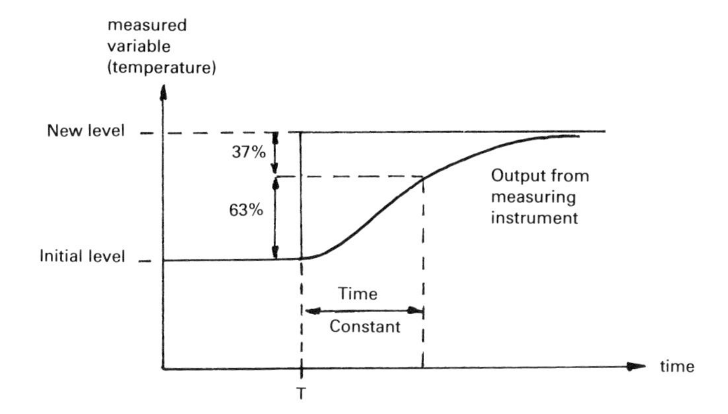

Figure 1.2 Dynamic response of a measuring instrument with a simple time constant when the measured variable increases sharply at time T.

Figure 1.2 shows the way in which a *first-order* (time lag) system responds to a sudden positive change to the physical variable – in this case, temperature. The output from the sensor increases exponentially towards the new level of the variable. Its rate of response is characterised by a *time constant*, defined as the time (in seconds) that the instrument takes to close the gap by 63%, as shown in the figure. This artifice has to be employed because, in theory, the instrument's output will approach the new level asymptotically, but never actually reach it.

Figure 1.3(a)–(c) illustrates three aspects of a *second-order*, mechanical mass/spring/damper system. The first of these three figures shows the effects of fluctuations in the monitored physical variable on the output of the system. The frequency ratio is defined as the frequency at which the physical variable is changing, divided by the undamped resonant frequency of the sensor, f_0. Therefore the value of 1 on the horizontal, logarithmic axis is the frequency at which changes in the variable would cause the sensor to resonate in the absence of sufficient damping. Ideally, the amplitude ratio should remain constant at 1.0, but when frequency components in the variable's change exceed $0.1\,f_0$ distortion of the sensor's output occurs, to a degree depending on the damping ratio. It can be seen from Fig. 1.3(a) that a damping ratio of a little over 0.6 maximises the uniform frequency response of the system.

Figure 1.3(b) shows the effect of different degrees of damping on the system when it is subjected to a short impulse. The level of damping at which it just ceases to resonate is described as the *critical damping* and the damping ratio in Fig. 1.3(a) is the fraction of the critical damping.

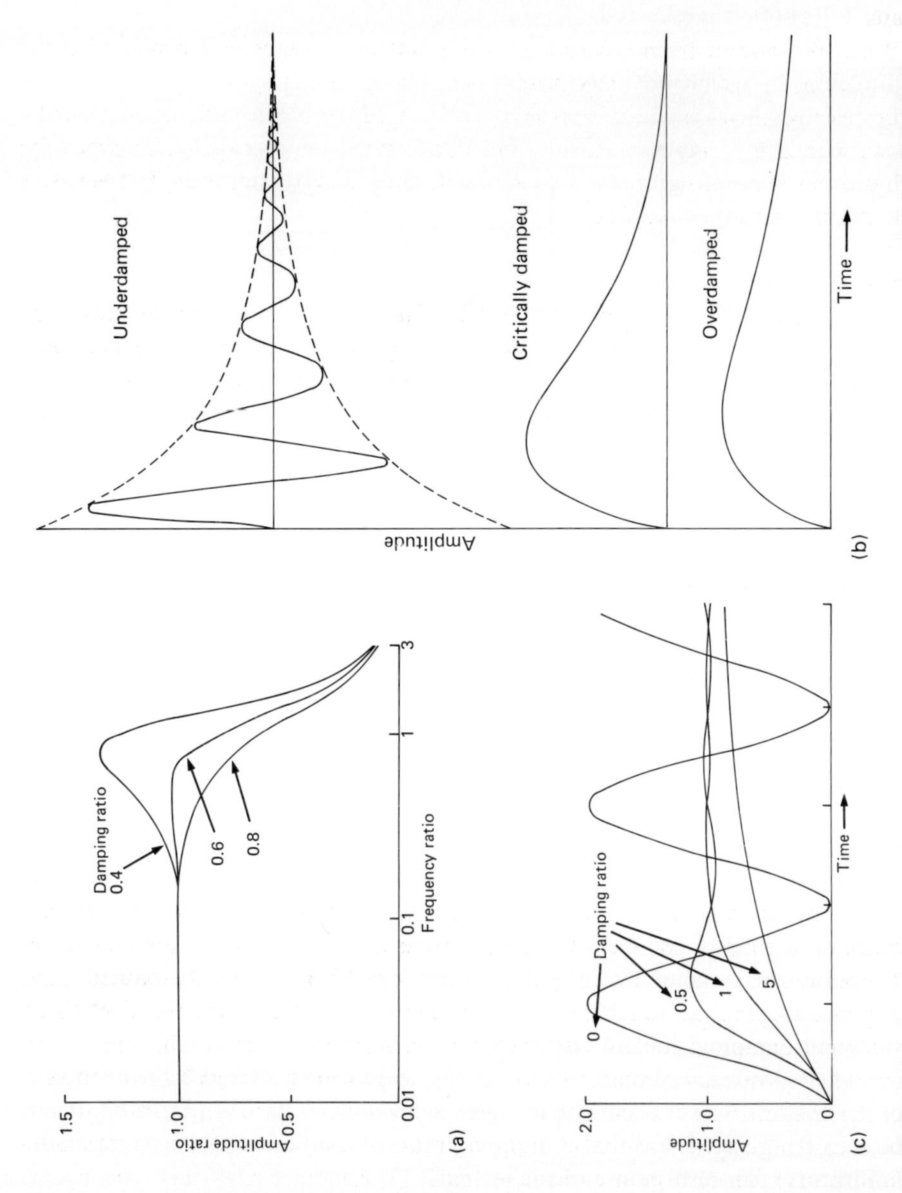

Figure 1.3 The dynamic characteristics of a mass/spring/damper system with different degrees of damping: (a) frequency response; (b) transient response; (c) step response.

Figure 1.3(c) shows the response of the system to a step change, equivalent to that in Fig. 1.2. It can be seen that critical damping produces a result similar to that of a first-order system.

Hysteresis

When the output from a sensor at any point in its range is dependent on the direction in which the monitored variable is changing (i.e., increasing or decreasing) then hysteresis is present. In effect, separate calibrations are needed – one with the variable increasing and another with it decreasing. Fortunately, hysteresis is not a significant problem with most sensors but some RH (relative humidity) sensors are prone to it.

Environmental effects

There are many adverse environmental influences on sensors which cannot be eliminated but which can be minimised by good design. Environmental temperature effects are the most common. Countermeasures include screening, to minimise heat exchange by radiation, convection and conduction; environmental control of the sensor's enclosure and the incorporation of temperature-compensating devices. Electrical sensors need to be protected against noise and surges on their supply lines and against external electrical and magnetic interference (EMI). Moisture is an insidious enemy of many types of sensors and it is one of the hardest environmental problems to eliminate.

1.3 Temperature sensors

Temperature is probably the physical property most commonly measured in agriculture and horticulture.

Expansion types

The familiar glass form of the liquid expansion thermometer rightly retains a place as a simple, compact and inexpensive means to monitor air and liquid temperatures, despite its fragility. Either hand-held or wall-mounted (and screened against heat radiation) it can serve in particular as an independent check that environmental control systems are performing as expected. This precaution (often overlooked) represents a small outlay to provide warning of the potentially costly malfunction that cannot be discounted in any system. A more expensive mercury-in-glass thermometer, with a calibration certificate issued by a standards laboratory, can also be a good investment. Thermometers of this type can be graduated to 0.1°C and calibrated to better than 0.05°C, although to obtain maximum accuracy they must be used in a uniform environment and if they are only partly immersed in a liquid an emergent stem correction should be applied. They are particularly suitable for checks on temperature probes such as those used in crop stores, which can be immersed in a stirred water bath, held at a range of temperatures, including the ice point. A large vacuum flask will serve as the container.

More robust (and more costly) mercury-in-steel or spirit-in-steel expansion thermometers provide reliable means for monitoring and recording temperature in harsher environments. The sensing bulb can be connected by a fine-bore capillary tube many metres in length to a remote indicating or recording dial, operated by a flexing Bourdon tube at the end of the capillary. Compensation is incorporated to counter the effect of temperature change on the diameter of the capillary bore and on the Bourdon tube. The accuracy of these thermometers (and of similar vapour pressure devices) can be within ±0.5°C.

Common bimetallic spiral thermometers, with or without a recording facility, have similar potential accuracy but these generate smaller forces than the liquid and vapour expansion types, therefore the mechanical linkage to the indicator or recording pen is less robust.

Thermocouples

The thermocouple is a widely-used industrial sensor, familiar as the safety device that detects the presence of the flame in gas- or oil-fired burners. It also has many uses in temperature measurement from below ambient temperatures up to about 1500°C, depending on the materials from which it is constructed. The principle of the thermocouple is simple, although understanding of its mechanism requires knowledge of solid-state physics. When two different metals (or metal alloys) form an electrical circuit a difference in temperature between the two joins causes a small current to flow round the circuit. If this loop is made open-circuit a small voltage between the ends can be measured and related to the temperature difference between the two junctions. If one junction is at a known reference temperature (say, the ice point) then the output is a function of the temperature to be measured. The effect (discovered by Seebeck in 1821) is due to the higher energy of the free charge carriers (electrons) in metals at higher temperatures and the difference between electron energy levels in the two metals.

[*Note* This is the reverse of the Peltier effect (discovered in 1834), which is the basis of the thermoelectric coolers now widely employed as refrigeration elements for components and equipment. In this case a current driven round the circuit causes heating at one junction and cooling at the other. These coolers employ semiconductor materials because the effect is much larger in them than in metals.]

The pairs of dissimilar metals commonly used for thermocouple measurements include copper–constantan (for temperatures up to about 300°C); nickel chromium–nickel aluminium (chromel–alumel) up to 1100°C and platinum–platinum rhodium up to 1500°C (see Table 1.2). Their advantages are that they can be of small size, with a resulting short time constant, but their output is non-linear, with the general form $V = aT + bT^2$, where T is the temperature difference in °C or K and the constants a and b depend on the metals. Their output is also low, at about 40 μV per °C for copper–constantan and chromel–alumel, and 10 μV per °C for

Table 1.2 Thermocouple types and ranges.

Type	Common names	Temp range (°C)
T	Copper–constantan	–250 to 400
J	Iron–constantan	–200 to 850
E	Nickel chromium–constantan or chromel–constantan	–200 to 850
K	Nickel chromium–nickel aluminium (NiCr–NiAl) or chromel–alumel (C–A) or (T_1–T_2)	–180 to 1100
—	Nickel 18% molybdenum–nickel	0 to 1300
N	Nicrosil–nisil	0 to 1300
S	Platinum 10% rhodium–platinum	0 to 1500
R	Platinum 13% rhodium–platinum	0 to 1500
B	Platinum 30% rhodium–platinum 6% rhodium	0 to 1600

platinum–platinum rhodium. When measurements are made at or near ambient temperature the cold junction's temperature must be constant to a fraction of a °C, which calls for a temperature-controlled enclosure equivalent to the ice-point reference flask already referred to, or an equally accurate automatic cold junction compensation circuit. Overall, the measuring system must be designed to avoid metal junctions and temperature variations which could generate unwanted thermoelectric voltages. Where extension leads are needed between the thermocouple hot junction and the meter compensation cables must be employed. These are wires that are either the same as those forming the thermocouple or are thermoelectrically similar to them.

These characteristics tend to limit the potential of the thermocouple as a tool for agricultural and horticultural use, but there is a wide range of commercial portable thermocouple meters suitable for spot monitoring of temperatures in these industries. They normally accommodate interchangeable, plug-in temperature probes of different types and forms. The type K thermocouple is generally favoured because of its wide range (Table 1.2) but type T probes are also available. The form of the probes varies from flat-ended units for surface temperature measurements to spears for measurement inside soft materials. Some are specifically designed for use with foodstuffs (meat and fruit). All are designed to meet national and/or international standards of accuracy. Table 1.3 provides calibration data to a British Standard specification, but the International Electrotechnical Commission's IEC584 is often quoted. Automatic cold junction stability is quoted at better than 0.1°C/°C ambient temperature and the ambient operating temperature of these instruments embraces the subzero temperatures that can be found in the food industry.

Electrical resistance types

A wider range of applications in the agricultural sector is open to the electrical resistance thermometer in its two main forms, i.e., the metal element and the semiconductor element. The metal element is usually made of platinum, platinum

Table 1.3 Thermocouple calibration data.

Symbol	Conductors	Specification	Class 2 – tolerances to BS 4937 Pt 20
T	Cu–Con	BS 4937 Pt 5	± 1.0°C or ± 0.75% up to 350°C max
J	Fe–Con	BS 4937 Pt 3	± 2.5°C or ± 0.75% up to 750°C max
E	NiCr–Con	BS 4937 Pt 6	± 2.5°C or ± 0.75% up to 900°C max
*K	NiCr–NiAl	BS 4937 Pt 4	± 2.5°C or ± 0.75% up to 1200°C max
R	PtRh–Pt	BS 4937 Pt 2	± 1.5°C or ± 0.25% up to 1600°C max

alloy or nickel wire, wound on an insulating former, although thin film versions of the platinum element are available. The last named are formed by depositing platinum onto a ceramic substrate, then cutting the deposit to the required conformation with a laser beam. However, the thin film element is far more susceptible to the self heating that occurs when a current passes through it during the measurement process. The wire elements are normally sheathed for their protection and the sheath is filled with a thermally conductive material, such as alumina powder, which improves the response time of the element and gives it mechanical strength.

Although platinum is more expensive than nickel and the platinum element's resistance changes only about half as much per °C the wire-wound platinum resistance thermometer (PRT) is widely used in industry. This is due to its wider range of operating temperatures and to its stable and reproducible characteristics, which make it a useful calibration instrument, as stated earlier. For comparison, the nickel element and the platinum foil element have a working range of –50°C to +200°C approximately, while the PRT is the internationally accepted standard for measurements between –200°C and +630°C.

As with the thermocouple, the metal resistance element's resistance/temperature characteristic is non-linear, being of the form $R_T = R_0\{1 + a(T - T_0) + b(T - T_0)^2\}$, where the suffixes T and 0 refer to the measured temperature T (°C) and a fixed reference temperature T_0 (usually 0°C) respectively. Again, a and b are constants appropriate to the metal used. Over moderate temperature spans the squared term can be ignored for practical purposes and the increase in resistance with temperature (the characteristic of metals) becomes $a(T - T_0)$. The a, or alpha, value of the PRT is internationally standardised as 0.003850 (IEC, 1983). This value is attained by alloying pure platinum (alpha value 0.003916) with other metals of the platinum group. That allows the production of closely matched elements, usually with a resistance of 100, 110 or 130 Ω at 0°C. The 100 Ω element is an international standard and is normally supplied with calibration to national versions of the standard, such as BS 1904 and DIN 43760. The element can also be supplied in two-, three- or four-wire versions – the latter two providing resistance compensation when long leads and the highest accuracy are required. Figure 1.4 shows the use of a four-wire element in the traditional Wheatstone bridge measuring system.

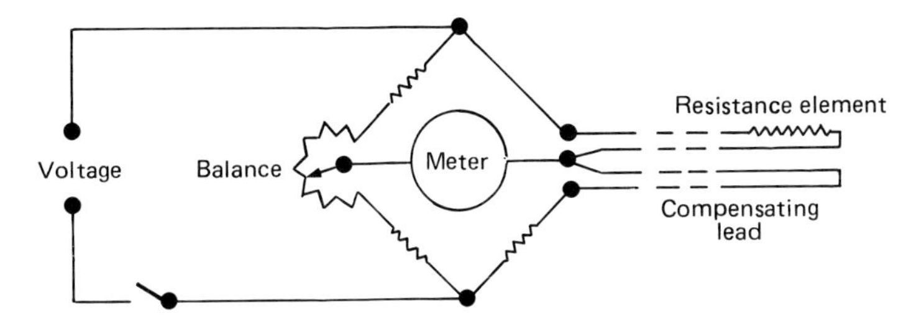

Figure 1.4 Resistance thermometer bridge, with automatic compensation for long leads.

The alternative, semiconductor resistance element, the thermistor, has found widespread application in agriculture and horticulture. Although it is less accurate than a PRT its cost is a fraction of the latter's and its performance is acceptable in many circumstances, given good design of the probe. In fact, quality control of thermistor production has made it possible to supply interchangeable units, akin to the thermocouple probes mentioned earlier. The temperature/resistance characteristics of these semiconductors are the reverse of those in metals, i.e., as their temperature increases their resistance decreases. The thermistor is therefore often referred to as an NTC (negative temperature coefficient) element. The relationship is exponential and is represented by $R_T = a[\exp(-b/T)]$ where a and b are constants for the material. However, in this case the T must be in the fundamental scale of temperature (K) since the relationship is not based on a temperature difference, as in the case of the thermocouple and the metal resistance thermometer, but on absolute temperature. The thermistor's normal working range is −80°C to +150°C.

The thermistor can be supplied as a small bead, with a correspondingly fast response time. An epoxy-encapsulated unit can have a time constant between 1 s and 10 s, depending on the convection loss to it surroundings. As an illustration of the resistance/temperature relationship above, an element with a resistance of 10 kΩ at 25°C could have a resistance of 40 kΩ at about −4°C and 2.5 kΩ at about 60°C. This is a much larger change than that of a metal element over the same temperature range and the higher resistance of the thermistors overall reduces both the influence of long leads and the problem of self heating. Interchangeability of these elements can be within ± 0.2°C over the range 0°C to 70°C, and within ±0.4°C from −40°C to 120°C. When traditional measuring instruments are employed the highly non-linear relationship between the thermistor's resistance and the corresponding temperature can be partly improved by connecting a temperature-insensitive resistance in parallel with it, although the sensitivity of the combination to temperature change is much reduced in consequence. An accuracy of ±1 to 2°C is attainable in this way. However, today's microprocessor-based instruments are capable of storing the sensor's R/T data and converting the measured resistance into the corresponding temperature automatically.

Silicon semiconductor temperature sensors with positive temperature coefficients are also available. These can be found in packaged integrated circuits mainly intended for temperature monitoring and control in other equipment, including provision of cold junction compensation in thermocouple-based thermometers. Their accuracy is usually between ±2°C and ±3 °C.

1.4 Humidity sensors

Fundamental standards

Measurement and control of air humidity is a common requirement in agriculture and horticulture but it is far from easy to monitor this environmental parameter. In fact, establishment of international and national standards has been a long, slow process, which is still developing. Much of the fundamental work was done at the USA's National Bureau of Standards (NBS) in the 1960s, while as recently as 1981 the NPL began a programme of work to establish a National Humidity Service for the UK, in response to representations from a wide sector of industry, including agriculture and the food industry. By 1986 the Laboratory was able to report on the commissioning of a gravimetric hygrometer to verify the performance of a humidity generator which would be the standard on which a calibration service would be based. Through this standard users and manufacturers would be able to obtain traceability of measurement hygrometers to UK National Standards. Procedures were under consideration for dealing with 'the wide range of secondary standards currently used in industry, additional to dew-point instruments, such as secondary-standard generators, salt solutions and bottled gas samples'. The calibration requirements of industry for lower accuracy calibration were being met by accreditation of laboratories within the National Accreditation Service, now the UKAS.

The gravimetric hygrometer

This uses an absorber to collect the water in a measured volume or mass of a gas. From the weight gained by the absorber the mass of water can be determined, to provide a measure of the absolute humidity of the air in g/m^3 or the mixing ratio in g/kg, respectively. The NPL has chosen to measure the mass of the gas, by liquefying it at very low temperature. Thereby the whole measurement is traceable to the SI base unit of mass. This method provides an accuracy of ±0.2% or better in measurement of the mixing ratio, at the 95% confidence level, for dew-point temperatures down to –40°C. Dew-point equivalents to the mixing ratios can be quoted to better than ±0.03°C at the same confidence level.

[*Note* At the dew point the gas is saturated with water vapour. Any further cooling will deposit dew (or frost). The mixing ratio under these conditions is termed the saturated mixing ratio. The ratio between the actual mixing ratio at a particular temperature and the saturated mixing ratio at the same temperature, expressed as a percentage, is termed the percentage saturation. This quantity is often equated to the relative humidity (RH) of a humid gas. Strictly, RH is defined as the percentage ratio of the actual vapour pressure of the water in the gas to its saturation vapour pressure at the same temperature and barometric pressure. In practice, the difference in these percentages is usually small enough to ignore.]

The dew-point generator

A dew-point generator provides a link between a gravimetric hygrometer and the calibration of industry standards. The NPL has two of these devices – one to calibrate dew-point hygrometers at temperatures down to −75°C and the other operating from 0°C to +82°C. The latter generates a gas stream saturated with water vapour by repeated circulation over a surface of water controlled at the desired dew-point temperature. The gas temperature is indicated by a platinum resistance thermometer and the process is controlled at a pressure of 105 kPa ± 0.02 kPa. (cf standard atmospheric pressure, 101.325 kPa) and gas can be supplied to an external device such as a gravimetric or other hygrometer at up to 3 litres/minute.

Uncertainties in the generated dew point, at the 95% confidence level, range from ±0.05°C at 0°C to ±0.07°C at +82°C. This standard is traceable to the SI base unit of temperature.

Two-pressure humidity generator

The NBS developed the two pressure system shown in Fig. 1.5 as a transfer standard. A sample of gas is first saturated at constant temperature t, and pressure, P_1, then expanded to a lower pressure, P_2, isothermally. Then the RH at P_2 and t is expressed as

$$\text{RH}\ (\%) = P_2/P_1 \times 100$$

This equation strictly applies to ideal gases, and a water vapour/air mixture deviates from this by a small amount. Work at the NBS showed that the correction to be applied to the values obtained by the above calculation varied from 0.25% RH at 10% to zero at 100% RH.

The dew-point sensor

Among the industry secondary standards listed by the NPL, the use of saturated and unsaturated salt solutions to provide calibration points at intervals over the RH range is effective but not particularly convenient. The same applies to the use of bottled gas samples. On the other hand, the dew-point meter has emerged as a useful, multipurpose instrument as a result of technological advances in its design over the past 25 years. For the last 10 years calibration of these meters over the temperature range −75°C to +82°C has been available from the NPL.

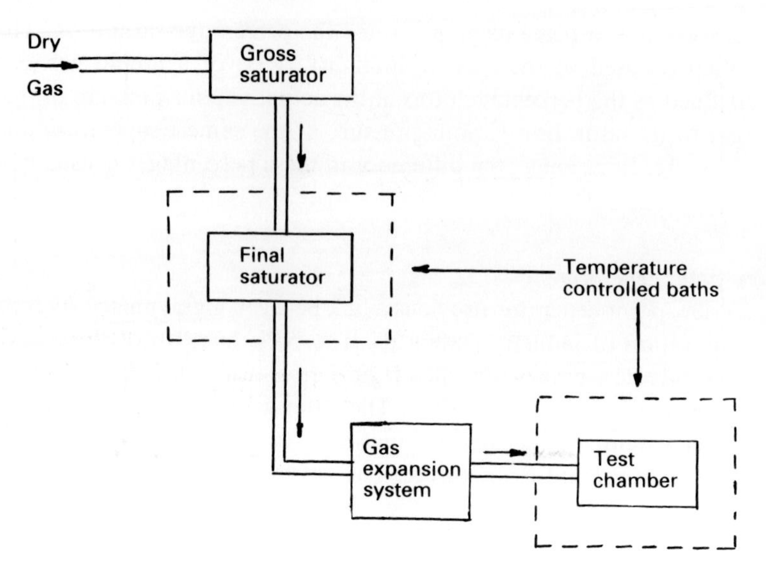

Figure 1.5 NBS two-pressure humidity generator.

The principle of dew-point measurement and its relationship to measurement of air humidity was first established 200 years ago. It involves slow cooling of a polished surface in contact with the moist air until the formation of dew on the surface is visible. The temperature of the surface at that time is taken as the saturation temperature of the air. The RH of the air at higher temperatures can then be deduced from psychrometric data. The dew-point sensor (the polished surface) must be clean and the temperature sensor must be in good thermal contact with it for maximal accuracy. The task of determining when the dew point has been reached can be left to a skilled operator but the dew-point meter replaces the operator by a photoelectric system.

Figure 1.6(a) is a diagrammatic representation of a fully automatic commercial meter. The sensor is a thin metallic mirror backed by a PRT which is coupled to the temperature indicator. The assembly is cooled by a Peltier effect device and the surface of the mirror is illuminated by a light-emitting diode (LED). Light reflected from the mirror is collected by a semiconductor photodetector. A second LED and photodetector form the other half of an electrical bridge which controls the gain (amplification) of a signal that, in turn, controls the power delivered to the Peltier cooler. The dynamics of the measuring system are adjusted to critical damping so that it quickly settles to a condition that maintains a thin dew layer. The rate at which the sample air passes over the mirror is not critical but minimum and maximum rates are generally recommended. The instrument shown in Fig. 1.6(a) has a display resolution (precision) of 0.1°C and an accuracy of ±0.3°C. Unlike many other methods of humidity sensing the dew-point method does not suffer from hysteresis. Because the mirror's temperature is measured by a PRT it is suitable as a laboratory reference instrument but it is also

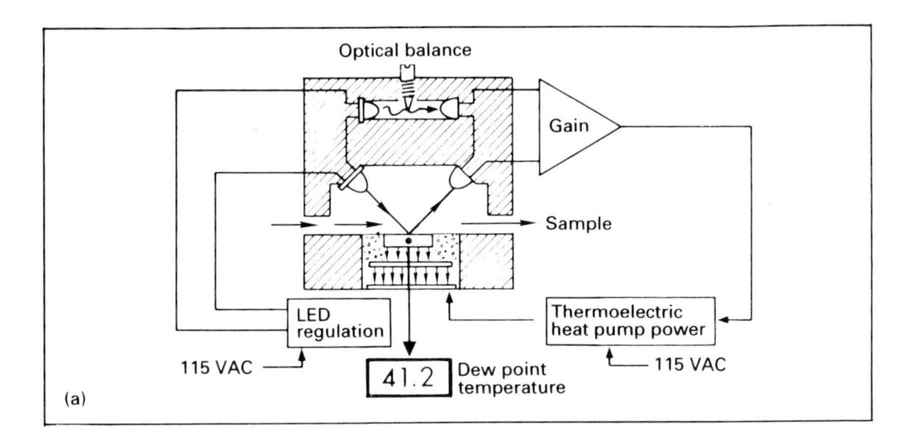

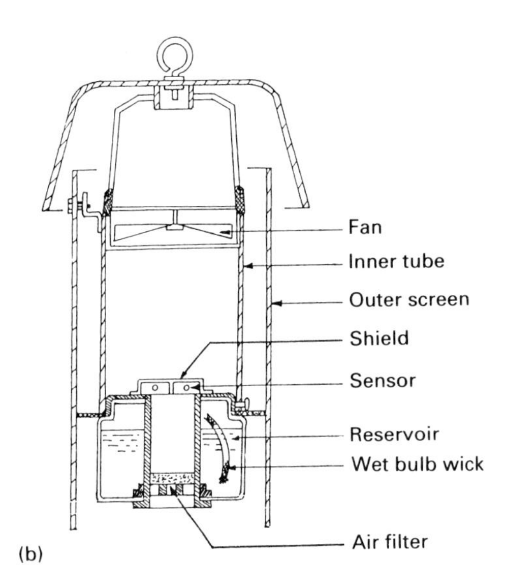

Figure 1.6 Humidity measurement: (a) dew-point hygrometer (General Eastern Instruments); (b) wet and dry bulb hygrometer (Victor Automation Systems).

suitable to many industrial environments where contamination of the mirror surface can occur, since it is designed for easy cleaning.

Absorption elements

Despite its ability to work in contaminated atmospheres the dew-point meter has not displaced simpler and less costly humidity sensors in agriculture and horticulture. These fall mainly into two classes – namely the absorption types and the wet and dry bulb hygrometer or psychrometer. The former range from the long-established mechanical hygrometer, in which the change in dimensions of a stretched hygroscopic material is converted to movement of a pointer or recorder pen, to electrical sensors in which two electrodes are separated by a film or layer of hygroscopic material and the resistance or capacitance between them is measured.

The mechanical types are sufficiently prone to drift, hysteresis, damage by contamination and limitations in their working range to make them less useful than other forms of hygrometers, although regularly serviced recording hair hygrometers were once widely used in environments free from extremes of humidity, temperature and pollutants. Under these conditions they were capable of maintaining an accuracy of better than ±5% RH in the range 15–85% RH.

The design, accuracy, hysteresis, drift and cost of electrical sensors vary widely but most claim an accuracy of about ±2% RH up to 80% RH and ±3% RH at the highest humidities. Time constants vary widely, too. For one capacitance sensor with a 1 μm (micron) thick dielectric polymer as the absorbing element, its response time in an air flow of 0.5 m/s is less than 5 s but for other capacitance elements it can be several minutes. Hysteresis can be up to 2% for a full excursion from 0 to 100% RH and back to 0, and drift can be ±1% RH over several hours at high levels of humidity, even with the best capacitance elements. Ambient temperature is unlikely to be a problem with these elements; many can operate down to −40°C and up to at least +60°C. However contamination in the atmosphere can seriously degrade this performance and even permanently damage the elements. Experience suggests that expectations of no better than ±5% RH are prudent in all but the cleanest environments.

The wet and dry bulb hygrometer

This also requires regular care and attention and cleanliness is essential to its operation. Figure 1.6(b) is a diagram of a continuous monitoring unit for environmental control. Its elements are essential for any form of the hygrometer. The most important feature is that the thermometer element that acts as the wet bulb (the *sensor* in the figure) should be surrounded by a tightly fitting and thoroughly clean wick, maintained in a wet state, by capillary action, with distilled water from the reservoir. The adjacent dry bulb must remain thoroughly dry. Then clean air from the environment to be monitored must be drawn over the two elements at an airspeed of about 3.5 m/s, to ensure maximum cooling of the wet bulb through uptake of water from the wick. This maximum *wet bulb depression* is a function of the RH of the air. The fan that draws air across the elements must be downstream of them, as shown, to avoid heating the incoming air and so reducing its RH. The air escapes from the annular gap in the radiation screen formed by the outer cylinder and the upper dome. The sensors used in this unit have to be as accurate as possible for control purposes, so they are PRTs, accurate to ±0.1°C. This allows humidity to be measured to about ±2% RH under the most favourable conditions.

1.5 Strain sensors

Many instruments for measuring mechanical force, in one form or another, employ strain gauges as the sensing elements. The relationship between force per unit area (stress) and the resulting change in shape or size of a body (strain) was studied experimentally by Robert Hooke (1679), who found that the strain was proportional to the stress in many materials over a considerable range (i.e., up to an elastic limit). This is known as Hooke's law. The stress/strain ratio in the elastic range is termed an elastic modulus and three moduli are defined, according to the nature of the strain:

- Longitudinal stress/fractional increase in length
- Tangential stress/angular deformation
- Compression or tensile stress/fractional change in volume

These are known respectively as Young's modulus (after Thomas Young, *c* 1800), the shear modulus and the bulk modulus. Young's contemporary, Poisson, established that a longitudinal tensile strain in a body results in a proportional lateral contraction, leading to Poisson's ratio, defined as the fractional decrease in width/longitudinal strain. All four of these elastic constants are inter-dependent, of course. Later (1856) Lord Kelvin reported that the electrical resistance of some conducting wires was proportional to their length and inversely proportional to their cross-sectional area; therefore a relationship existed between electrical resistance and mechanical strain in those materials.

Metal strain gauges

The development of the aircraft industry, with its need for measuring and analysing rapidly varying stresses in aircraft structures, led to the development of the modern foil strain gauge, manufactured by the printed circuit process, which produces inexpensive gauges to close tolerances.

Figure 1.7(a) shows a common form of the foil strain gauge, 2.5 μm thick. The insulating backing (25 μm thick) is bonded to the body in which stresses are to be measured. Its function is to ensure that the gauge reproduces any strain in the body as faithfully as possible while isolating it electrically from the body. The two printed arrowheads on the left and right of the gauge in the figure define the longitudinal axis of the gauge, while the other two arrowheads define its transverse axis. The four short bars indicate 45% axes relative to the first two. The shape and thickness of the sensitive area (the grid and its end loops) are such that the gauge is flexible and able to respond to changes in the stressed body, while having sufficient current-carrying capacity to provide a satisfactory signal. Stress concentrations in the grid are minimised by the wider sections, while the wide end loops help to reduce transverse sensitivity. The grid bonding area also minimises hysteresis and creep.

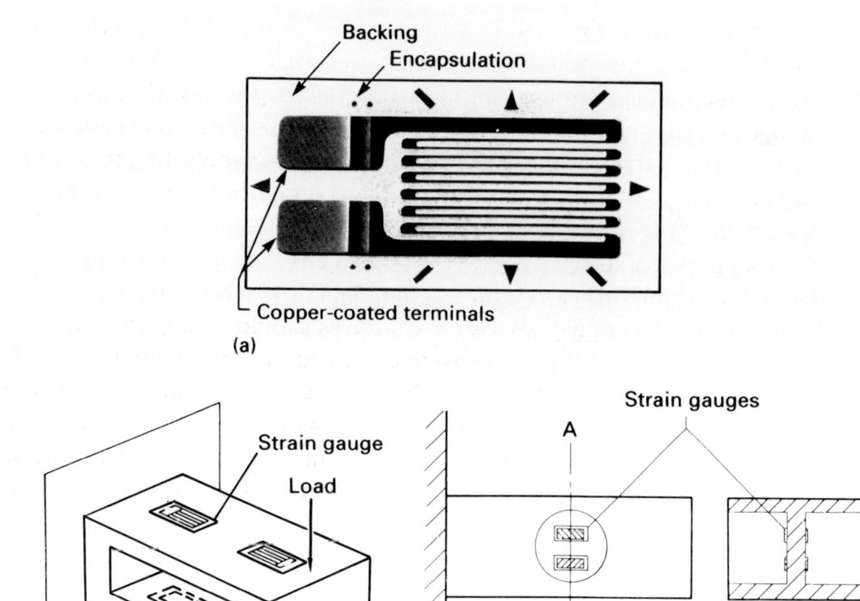

Figure 1.7 Load measurement. (a) Etched foil strain gauge, with insulating backing and protective coating over the strain-sensitive grid. Grid size approximately 10 mm (length) × 5 mm (width). Strain is applied in the longitudinal direction. (Welwyn Strain Measurement). (b) Double cantilever load cell. (c) Shear web load cell (Griffith Elder).

Gauges are made in several stock sizes and in varied combinations. For example, three of the type shown in Fig. 1.7(a) can be *stacked* (superimposed) using the 45° axis markers to produce a combination which simultaneously measures strain in three directions at 45° angular separation at a particular point on the body. Special grid patterns with circular symmetry are produced for measurement of strain in circular diaphragms. However, all have common requirements. The relationship between their change of resistance and strain should be as linear as possible and as high as possible, while their temperature sensitivity should be as low as possible. In addition their resistivity should be high, in the interests of measurement sensitivity. The relationship between the change in resistance that accompanies a change in strain is called the *gauge factor*. For common gauges this is about 2.1. These gauges are made of copper and nickel alloys with resistances of 120 Ω and upwards and temperature coefficients of resistance of 2 to 8×10^{-5} per °C. Constantan has excellent properties in this respect but is not reliable above 65°C, whereas a nickel/chrome alloy known as Karma can be used up to 300°C. The temperature coefficients of both alloys can be adjusted in the fabrication process to match the temperature coefficient of the material to which they will be bonded, over a range of temperatures. This self-temperature compensation is a valuable property when the gauges are used in transducers. An important limitation of all strain gauges is their maximum strain range, which is normally of the order of 2 or 3%. They also have a fatigue life under cyclic strain, which is usually quoted by manufacturers in terms of the number of cycles guaranteed at a given maximum strain level. These are of the order of 10^6 or 10^7 cycles.

Fixing gauges to the body concerned requires the utmost care and not a little skill, particularly if an assembly is the basis of a transducer. The surface of the body must be scrupulously clean; adhesives should be applied thinly (less than 10 μm) and without voids, to ensure close and uniform bonding of the backing; damage to the gauge must be avoided when connecting wires are soldered to its terminals and waterproofing of the whole assembly must be thorough. Every effort must be made to exclude moisture from the gauge before, during and after assembly, including the use of connecting wires selected for their waterproof properties. Most failures of strain gauges are caused by the ingress of moisture, which causes drift and corrosion. Training in affixing these gauges is therefore highly advisable for the beginner.

Since it is a resistance element, like the PRT, the strain gauge can be made part of a bridge network such as that shown in Fig. 1.4, using an adjacent dummy (non-strained) gauge in the compensating lead to provide temperature compensation. However, if the gauge is of the self-temperature compensation type the dummy gauge is unnecessary. In fact, several different bridge networks are employed with these sensors.

Strain gauges bonded to deformable elements are the basis of many types of transducer, particularly for pressure and load measurement. In load cells gauges are often mounted in pairs, in such a way that the applied load subjects one to extension and the other to compression. This arrangement increases the sensitivity of the transducer. The shapes of the deformable elements vary, from single cantilever beams of simple or complex cross-section to folded shapes with outputs of high linearity. Some have outputs which are conveniently insensitive to the position of loading (Figs 1.7(b) and (c)).The shear web load cell is highly resistant to side loads, which makes it particularly suitable for vehicle weighing (see Chapter 5).

Silicon strain gauges

Within the past decade pressure transducers incorporating silicon strain sensors have been available commercially. The discovery that silicon and some other semiconductors have a large piezoelectric characteristic was made in the mid 1950s. The gauge factor of a silicon strain can be over 100 (positive or negative) – i.e., 50 times greater than that of the common metal alloy gauges. That high output made it possible to market relatively simple and inexpensive instruments, although in their original form (strips cut from single crystals of silicon) they suffered from relative lack of stability. This was largely due to the nature of the adhesive used, including its differential expansion relative to the material to which the gauges were bonded. They also had a more limited upper strain limit (less than $\frac{1}{2}$%) Subsequently, improved sensors have been made by diffusing boron into a silicon wafer to create the required gauge pattern. The wafer itself is a pressure diaphragm. These sensors have very low pressure and temperature hysteresis but they suffer from temperature-dependent leakage currents within the gauge structure and from the need to compensate for differential thermal expansion of the component materials in the diaphragm assembly. A further development has been based on chemical vapour deposition. By this technique an insulating layer of silicon dioxide is deposited onto a stainless steel diaphragm. Then a layer of polysilicon is added before it is doped and etched to create the gauge pattern. Finally the whole assembly has a passivation layer deposited on it. This process takes advantage of the high output from silicon gauges without incurring the high costs of some other silicon-based pressure sensors.

1.6 Piezoelectric sensors

It has been known for most of the twentieth century that some classes of dielectric crystals possess important electromechanical properties. Under mechanical stress applied in particular directions separation of internal electric charge takes place so that some regions become positively charged and others negatively charged. This is known as the piezoelectric effect. Quartz was found to produce the effect both in compression and shear. Compressing a quartz crystal perpendicular to an axis of symmetry causes charges of opposite sign to appear at the two compressed faces. Extension in the same direction produces the same effect but with the polarity at the faces reversed. Conversely, placing the crystal in an electric field expands or contracts it according to the sense of the field. When the field oscillates rapidly the damping of the vibrations in the crystal is very low and a very sharp resonant frequency can occur. The resonant frequency also has a very low temperature coefficient. These are the characteristics which make the quartz crystal oscillator such a simple yet highly stable frequency/time standard.

The piezoelectric effect can be exhibited in a number of ceramic compounds which are used both to generate and measure vibrations of many kinds, from acoustic frequencies up to those in the MHz range. They have found application in agriculture and related industries in several ways, to be described in later chapters. The effect can also be found in some polymers that can be made into membrane sensors. One of these is polyvinylidene fluoride (PVdF), which has been utilised in hydrophones, operating in the MHz frequency range, for medical studies of ultrasonic fields propagating in water. Its strong piezoelectric coefficient and its acoustic impedance relative to that of water give it an advantage over ceramic hydrophones in that context. Metal leads evaporated onto both sides of the thin membrane define the active area of the device. Piezoelectric sensitivity is then induced in that area by a process known as poling. In this process, which is applied to other fabricated dielectric sensors, the material is raised to a high temperature and a voltage applied between the electrodes to create an electric field within it. This has the effect of aligning the electric dipoles in the material. The element is then allowed to cool, with the field still applied, leaving the material in a permanently polarised state. Pressure changes thereafter affect the captive surface charges in proportion to the material's piezoelectric coefficient. At any time, the surface charge, Q, converts to the measured voltage, V, through the relationship for a capacitative element, $Q = VC$, where C is the element's capacitance, arising from its relative permeability, or dielectric constant.

The hydrophones are of small membrane area but the material can be made in sheets for a wide range of potential applications, including some in the agriculture and food sectors. The film is tough and chemically inert, thermally stable and moisture resistant.

1.7 Pyroelectric sensors

Polyvinylidene fluoride film also has pyroelectric properties, which it shares with single-crystal triglycine sulphate (TGS) elements and with ferroelectric ceramics (after poling in all cases). These sensors are widely used for detection and measurement of infra-red radiation. As with piezoelectric materials, they have a characteristic coefficient, the pyroelectric coefficient. This has a non-linear relationship with their temperature, dropping to zero at a temperature known as their Curie point.

1.8 Radiation sensors

The spectrum of electromagnetic radiation (radiated energy) covers an enormous range of wavelengths and their corresponding frequencies. [*Note* frequency (Hz) = speed of light (3×10^8m/s)/wavelength (m). Wavelengths, rather than frequencies, are usually quoted over most of the range.] Table 1.4 summarises that range and the names given to subregions of it. The boundaries between regions are not hard and fast. They reflect physical processes in some instances and technology in others. For example, gamma rays originate from nuclear changes; the response of the human eye defines the narrow visible light band and the longer wavelengths chart different forms of radio transmission.

The table also draws attention to the well-known *wave/particle* duality of light and radiation in the other regions of the spectrum by showing the correspondence between wavelength and the energy carried by each *particle* (photon). The 10^6-fold increase in photon energy from the near infra-red to the gamma-ray end of the spectrum can be seen.

All of these characteristics have an influence on the sensors employed for

Table 1.4 The electromagnetic spectrum.

Spectral region (approx)	Wavelength	Frequency	Photon energy
Gamma rays	1 pm		1.2 MeV
X-rays	10 pm		120 keV
	0.1 nm		12 keV
	1 nm		1.2 keV
Ultraviolet	10 nm		120 eV
Visible	500 nm		12 eV
Near infra-red	1 μm	300 THz	1.2 eV
Infra-red	100 μm	3 THz	
Microwave	0.5 mm	600 GHz	
(Satellites/Doppler speed meters)	30 mm	10 GHz	
	300 mm	1 GHz	
TV/VHF	3 m	100 Mz	
MW radio	300 m	1 Mz	
LW radio	1 km	300 kHz	

(1) Frequency (Hz) = speed of light (m/s)/wavelength (m)
= 3×10^8/wavelength
(2) Photon energy (J) = h × frequency (Hz)
where h = Planck's constant
= 6.6×10^{-34} J s
From 1 and 2: J = 2×10^{-25}/wavelength
(3) electron volts (eV) = 1.6×10^{-19} J
(4) SI prefixes: p = 10^{-12} k = 10^3
n = 10^{-9} M = 10^6
μ = 10^{-6} G = 10^9
m = 10^{-3} T = 10^{12}

monitoring radiation in particular regions of the spectrum. At high photon energies, where photon flux tends to be lower, photon counting equipment prevails. The widely used scintillation counter for gamma ray work employs a phosphor such as sodium iodide as a sensor. This reacts to the photon flux by producing light flashes (one per photon) which are detected by a closely-coupled photomultiplier tube, with a photocathode responsive to the light spectrum of the phosphor. The pulse output from the photomultiplier provides a photon count and the size of the amplified pulses can be used to determine the energy of the photons. In the infra-red region there is a selection of broadband radiation sensors which measure *total* radiation in various ways – some by direct pyroelectric effect, others by capturing the energy in an absorbing receiver and measuring the receiver's temperature with thermometer elements. In other regions – chiefly in the visible range – several different types of photosensitive sensor are in use. Their spectral range and sensitivity are varied, as is their field of application.

More details on all of these sensors can be found at appropriate points in later chapters.

1.9 Position and displacement sensors

The general characteristics of a variety of sensors for measurement of position and distance are gathered here. Position fixing by general surveying methods is not covered in this book but the use of satellite reference data and other means for navigation in fields is included in Chapter 3.

The linear variable differential transformer (LVDT)

Here the characteristics of a widely used sensor for measurement of relative position and linear displacement on a small scale are introduced. This is the LVDT, shown diagrammatically in Fig. 1.8(a). This device is essentially a transformer with a single primary coil and two identical secondary coils disposed on either side of it, wound on the same cylindrical bobbin. A movable ferromagnetic core is free to move axially within the unit. The distance between an object and a reference frame can be measured by attaching one to the central core and the other to the bobbin assembly. The primary coil is energised by alternating current, which can be at mains frequency but is sometimes higher, depending on the speed of response required of the measurement system.

Referring to Fig. 1.8(b), when the core is central and the two secondary coils are connected so that their output is in opposition, the a.c. output voltage is at a null – ideally zero but in practice small – as shown. Displacement of the core to either side of the null position creates an output voltage which increases linearly throughout the working range. The voltage provides no indication of the direction of the movement; that can only be determined by reference to its phase, which changes by 180° on either side of the null position. This problem can be overcome, in essence, by connecting the secondary coils as shown in Fig. 1.8(c) and rectifying their output to d.c. Their combined output is then a linear function of the core's displacement, which changes polarity at the null point. Refinements to this basic circuit reduce the sensor's temperature dependence, which can lie between 0.05% and 0.005% per °C. Means for reducing the effects of drift in the null position can also be incorporated in the circuit. However, the LVDT's output is highly reproducible, since it has no mechanical hysteresis. It can have a non-linearity of ±0.5% or better and a sensitivity of $\frac{1}{2}$V per μm. It is also available with a range of displacements up to 500 mm, but for longer displacements the primary coil is distributed along the bobbin.

The coil assembly as a whole is very robust, since it is in a ferromagnetic case, which also provides magnetic and electrostatic screening. Cleanliness of the

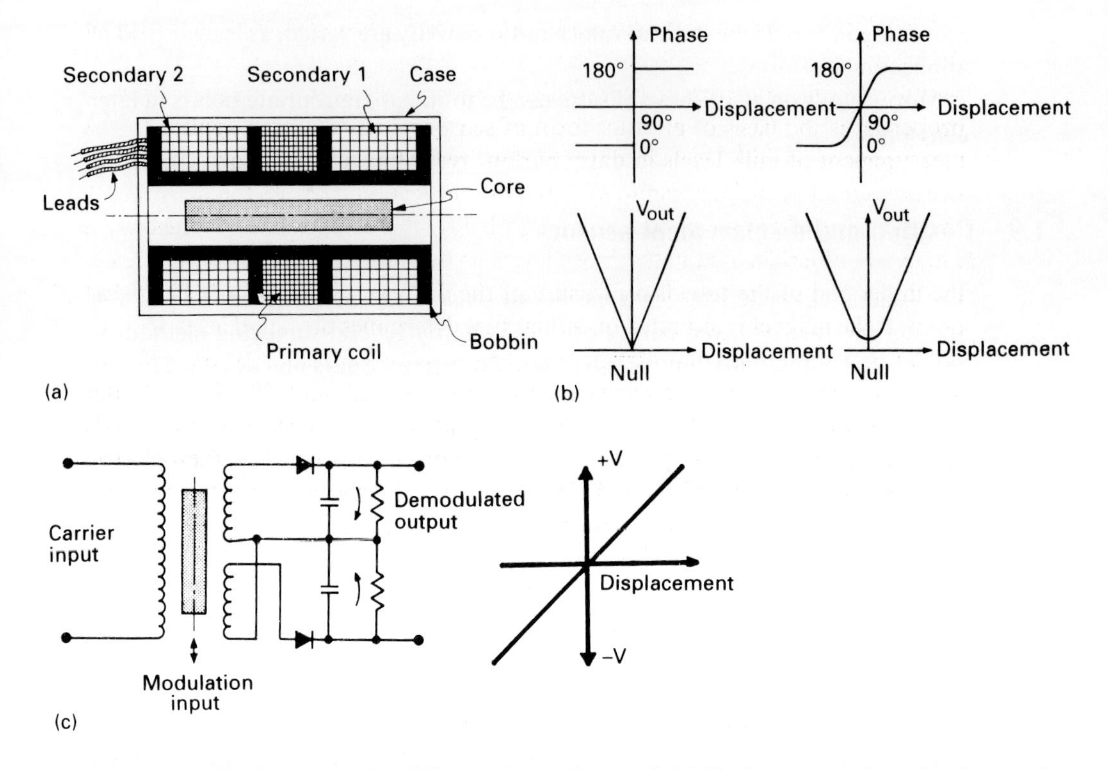

Figure 1.8 The linear variable differential transformer (LVDT): (a) construction; (b) phase and voltage changes at the null point; (c) bi-directional output.

emergent extension of the core is necessary, of course, to allow it to move freely. However, it is possible to operate the sensor with the core inside an enclosure and the coil assembly outside it, so long as the intervening boundary material does not interfere with the magnetic coupling between them. Whether this measure is taken or not, the LVDT is less prone to abrasion damage than the simpler electrical resistance potentiometer for measurement of varying distance.

A rotary version of the LVDT (the RVIT) was developed in the early 1990s, to provide a friction- and hysteresis-free means for measuring angular displacement and rotary motion, over a range of $\pm 60°$, with an accuracy of $\pm 0.5\%$. The coils are in flat, printed-circuit form and the core is a semicircular disc mounted on a central spindle. Apart from the core, the whole assembly, including the associated electronics, is on a printed-circuit board.

Level sensors

Surface levels of solids and liquids can be measured with a variety of sensors – inductive, resistive, capacitative, magnetic, ultrasonic and optical. Surface contact probes utilise LVDT and potentiometric resistance elements (linear or rotary). These can also be employed for liquid level measurement, with the use of floats. Linear and rotary encoders can be used with the probes to provide a digital output for subsequent data processing. A tube of metal with magnetostrictive properties is the basis of another form of sensor which has been applied to the measurement of milk levels in dairy parlour recording jars, inter alia. In operation, a current pulse is transmitted down the vertical tube. When the pulse enters the magnetic field of an external magnet, which can slide up and down the tube, a torsional pulse is created in the tube. The time that the latter pulse takes to reach the upper end of the tube is a measure of the distance of the magnet from that point. If the magnet is mounted on a float that determines the liquid level. Ranges of 3 m or more are attainable with this sensor. Non-linearity is of the order of 0.05%; the temperature coefficient is extremely low at about 0.002% per °C and the temperature range is 0–65°C. Ultrasonic pulse-echo level detectors for liquids and slurries have even longer ranges, from about 0.25 m to 10 m. Piezoelectric elements provide the pulses and detect the return echoes, while temperature sensors provide the data for correction of the pulse-echo time interval to take account of the variation of the speed of sound in air at different temperatures. After that correction, range accuracies of ±0.25% are attainable. Equally, ultrasonic ranging can be employed to measure height above ground surface level, and has been used in this way in agriculture (see later chapters).

Another level-sensing system, usable with liquids and slurries, employs a corrosion resistant probe, forming part of a radio frequency (RF) bridge circuit, which measures the electrical admittance (inverse impedance) between this sensor and the surrounding earth potential. Once calibrated, the associated electronic circuit is designed to ignore build-up of material on the probe. The probe can be many metres in length.

Measurement of hydrostatic pressure is proportional to the height of a liquid column; therefore a pressure sensor in the base of the container will provide a level determination, regardless of the vessel's shape and the liquid's viscosity.

Proximity sensors

At shorter range, inductive proximity sensors measure the range to metal surfaces at distances between 2 and 10 mm, with ±0.25% non-linearity and temperature drift of about ±0.1% per °C. The application of these devices for measurement is rather specialised but as proximity switches they have found agricultural applications as means to measure rotation (step-wise) in association with a rotating ferromagnetic, toothed wheel or ring. More recently the smaller and cheaper Hall effect device has been used for the same purpose. This effect (discovered by E.H. Hall in 1879) occurs in conductors carrying a current and in a magnetic field transverse to the current direction. A potential difference then appears across the conductor in the third axis, perpendicular to both, as a result of a drift of the conductor's charge carriers in that direction. This property is pronounced in some semiconductors. Therefore, if a rotating wheel carries a magnet or magnets which pass close to a fixed Hall effect module it produces an output pulse at each passage.

However, greater understanding of the phenomenon of magnetoresistance, discovered by Lord Kelvin in 1856, led to the giant magnetoresistance (GMR) element in 1986. Magnetoresistance, as measured by Kelvin, was a slight effect on the resistance of conductors placed in a magnetic field, but the GMR element is constructed of multiple layers of alternately polarised magnetic film, interleaved with non-magnetic layers in such a way that the current electrons in the material are seriously impeded until the element is placed in a magnetic field. At that point the resistance drops sharply. A unit with a much higher output than the Hall effect device is created in this way. It also operates at much lower magnetic field strengths. Linear and rotary position and rotation sensing are only part of the element's capabilities.

In the present context the role of the capacitance sensor is mainly restricted to that of a level switch. In that form it has found agricultural application in materials handling operations (see later chapters). Similarly, standard photo-sensors serve in many devices designed to detect the position or numbers of objects, through reflection or interruption of light beams.

1.10 Velocity sensors

The velocity (i.e., speed and direction) of solid bodies can be derived from the rate of change of the output from all of the above analogue, digital and pulse-echo sensors, with the reservation that the last two types are inherently position-sampling devices, which can only provide sampled data on the velocity.

The Doppler speed (or radar) meter, based on the effect noted by C. Doppler (1842), measures the apparent change in frequency of sound or electromagnetic radiation emitted from a source when the source and the receiver are moving relative to one another. Figure 1.9 illustrates the effect in its simplest form. In the diagram S is the source of constant frequency, f, and O is the receiver (or observer). AS is the wavelength (λ) of the radiation. Assuming that the speed of propagation of the radiation through the intervening medium is constant (c m/s) the observer receives f wavecrests per second. If the source then recedes at a constant speed v_s m/s the new wavelength is $\lambda_s = \lambda + v_s/f$ and the frequency of the signal received at O is then $f_s = c/\lambda_s$. Combining these two equations with $f = c/\lambda$ yields:

$$f_s = f\frac{c}{c + v_s}$$

i.e. the apparent frequency at O is reduced. If S moves towards O (v_s is negative) the apparent frequency increases. The result is similar if S is stationary and O is in motion, or if both are in motion either towards or apart from each other.

[*Notes* When the source is a sound generator and the medium is air, the speed of propagation, c, can vary substantially with temperature and pressure.

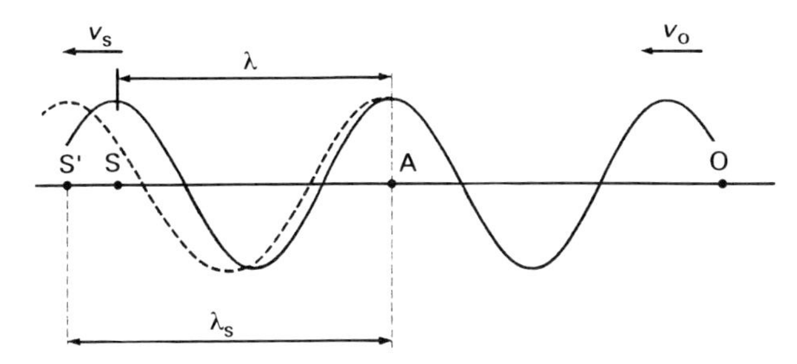

Figure 1.9 The Doppler effect.

When c is the speed of propagation of an electromagnetic wave (3×10^8 m/s) in most cases v_O and v_S are very much smaller than c. Therefore the fractional changes above are also very small.]

1.11 Liquid flow sensors

Pipes

Doppler measurements at ultrasonic frequencies are employed to measure liquid flow in pipes. The pulse-echo method depends on the presence of particulate matter or air bubbles entrained in the moving fluid to provide the necessary return signal to the sensor. The method has the advantage that it can be non-invasive. The sensor is strapped firmly to the outside of the pipe, using a suitable grease or resin to ensure good acoustic transmission (i.e. no reflections at the sensor/pipe interface). Velocities down to 0.3 m/s can be measured in this way with accuracies better than $\pm 5\%$.

For liquids with little or no entrained material two ultrasonic sensors can be employed as transit time detectors (Fig. 1.10). Each acts in turn as pulse generator and receiver. The difference between the upstream and downstream transmission times over the known distance between the sensors is a linear function of the flow rate, although compensation is necessary to correct for changes in the velocity of sound with the temperature of the fluid. Modifications of this technique include a clamp-on sensor assembly which avoids the need for critical alignment of the two sensors in Fig. 1.10 by creating resonance along a length of the tube, so allowing signal reception over a wide angle. By this means, density variations due to temperature changes or compositional changes of the fluid can be monitored, in addition to its flow rate (bidirectionally). This instrument can respond to flow rates from 0.3 mm/s upwards, with non-linearity of 0.1% accuracy of 0.25% of actual flow and a frequency response up to 10 Hz.

Other methods of sensing liquid flow through pipes are shown in Table 1.5, which lists some of their main characteristics. All are intrusive, monodirectional and they must be matched to a given pipe size.

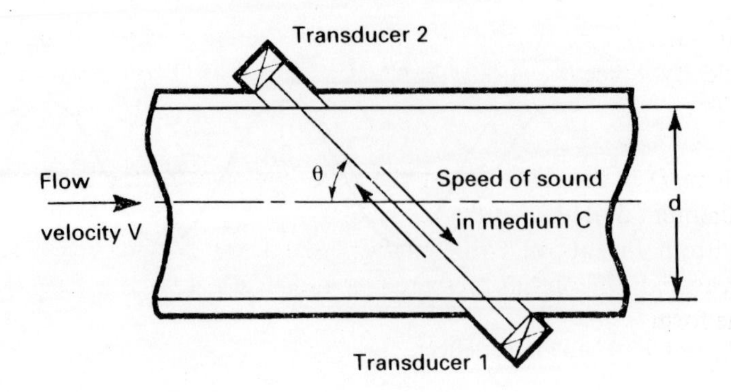

Figure 1.10 Doppler flow meter.

The differential pressure sensors (orifice plate and Venturi constriction) are industry standards, with virtues of simplicity and of reliability if contamination of thc fluid is avoided. They make use of the relationship between the static pressure in a fluid and its kinetic energy. That relationship was established by Bernoulli (*c* 1740) for non-viscous, incompressible fluids in steady (streamline) flow. Bernoulli's equation is of the form:

Static pressure + kinetic energy = constant

The kinetic energy is given by $\frac{1}{2}\rho v^2$, where ρ is the constant density of the incompressible fluid and v is its velocity. Therefore, when the fluid meets a restriction, such as an orifice plate or Venturi constriction, the increase in its velocity is compensated by a decrease in its static pressure, as follows:

$$p_u + \tfrac{1}{2}v_u^2 = p_d + \tfrac{1}{2}v_d^2$$

where p is the static pressure and the suffices u and d refer to upstream and downstream of the restriction, respectively. Measurement of the differential

Table 1.5 Calibration of pipe flow meters.

Performance	Orifice plate	Venturi	Vortex shedding	Magmeter	Turbine
Accuracy	2% of full scale	1% of full scale	2% of actual flow	1% of full scale	0.25% of actual flow
Turn-down ratio	4:1	4:1	20:1	10:1	20:1
Linearity	5%	2%	2%	1%	2%
Head loss	High	Moderate	High	Zero	Moderate
Limitation	Corrosive liquids	Susceptible to contamination	Changes in viscosity	Susceptible to contamination	Changes in viscosity

pressure, $p_u - p_d$, by any suitable means, then provides a measure of the change in velocity. Since the volume flow, Q (m^3/s), in an incompressible fluid is constant:

$$Q = Dv_u = Av_d$$

where D is the diameter of the tube upstream of the restriction and A is the minimum diameter of the restriction.

From the above two equations the form of the relationship between the measured differential pressure and the required volume flow can be reduced to the form

$$Q = \mathrm{k}A\sqrt{p_u - p_d}$$

where k is a constant. This is the general, non-linear response of the differential pressure sensor.

From the first two equations above it is not difficult to deduce an equation for k, under the idealised conditions in which the Bernoulli equation applies. However, in practice the value of this constant, known as the discharge coefficient, has to be determined empirically. Fortunately, values determined in this way hold over wide ranges of flow and standardised designs provide sensors with known and reliable discharge coefficients, for both liquids and gases.

In vortex meters, a piped liquid or gas flows around a central, non-streamlined bluff object and separates from it, downstream, in a turbulent wake. Above a minimum velocity the turbulence forms a regular pattern of vortices, produced at a rate proportional to the rate of fluid flow. These can be detected by several means, including the modulation of a transverse ultrasonic beam. The primary output is therefore a pulse train, with a frequency proportional to the flow rate. These devices are also used with both liquids and gases.

The magnetic flow meter is used with conductive fluids, which flow through the magnetic field created by field coils external to the pipe. As with Hall effect devices, a voltage is generated in a direction perpendicular to the conductor (the fluid's direction) and to the magnetic field. This is measured by electrodes on opposite sides of the pipe and at right angles to the field coils. It is proportional to the velocity of the fluid with only a small non-linearity, as shown in Table 1.5.

Turbine meters for piped liquids exist in many forms but their output is normally the same, i.e., a pulse train derived from a proximity sensor (usually magnetic) which senses the passage of the blades. They are sometimes placed in a bypass channel which bridges an orifice plate. In that way they carry a constant fraction of the total flow and become high range meters. In addition to the dependence of their output on the viscosity of the fluid the quality of their bearings is of paramount importance. These must be of low friction and corrosion resistant. Sapphire cups and stainless steel spindles are commonly employed.

Open channel

Open channel measurements of fluid flow employ some of the above pipe flow devices. For example, pipe flow conditions can be created by incorporating a piped section with an orifice plate in the centre of a meter box which has an upstream dropboard and a downstream weir, to ensure complete filling of the pipe. The length of pipe must be sufficient to establish steady flow before the orifice is reached. Alternatively, open flow metering can be done with a sharp-edged weir of rectangular or V-notch section, the latter giving better sensitivity at low flow rates. The measured quantity is the height of the water just upstream of the weir relative to its lower boundary. In the case of the V-notch this boundary is the apex of the notch. Weir geometry is standardised, like that of orifices, and scaling of the device is possible from standard formulae.

The weir, like the orifice plate, results in a substantial head loss in the water discharge system, and is prone to error due to silting action, therefore an equivalent of the Venturi tube may be used in open channels. This Venturi flume also provides a measure of flow rate in terms of water head.

With both weirs and flumes the measurement of head is usually done in a side chamber, linked by pipe to the main flow channel. Float-operated indicators, recorders and integrators have traditional use at this point but ultrasonic height sensors have been introduced for these situations, as has the pollution-tolerant RF admittance sensor mentioned earlier. The latter is in strip form for this application and it is slotted into the side wall of a flume. Both of these devices have the advantage that their outputs are referred to electronic equipment capable of automatic conversion of their height measurement into flow rates, via stored information on the shapes and dimensions of both weirs and flumes.

1.12 Gas flow sensors

Sensors commonly used for velocity measurement in air and other gases include the pitot tube (H. Pitot, *c* 1730), which – in its pitot-static form (Fig. 1.11) – can be used as a standard for calibration of other gas flow meters if it is constructed to established standards. Its narrow-bore tube, with a rounded or tapered tip, faces the direction of gas flow. The pressure in the tube is then the sum of the static and dynamic pressures already discussed in relation to differential pressure sensors. The outer tube with peripheral holes delivers the static pressure only to the measuring system, which can then determine the dynamic (velocity-related) component from the differential head, Δp (i.e., the difference between the inner and outer pressures). Then, by reference to ambient temperature T (converted to fundamental temperature, K) and pressure P, the gas velocity can be derived from a relationship in the form

$$v = \mathrm{a}\sqrt{\mathrm{b}T/P} \times \sqrt{\Delta p}$$

where a and b are constants. Here T and P provide corrections for changes in the density of the gas, which cannot be treated as a constant, in general.

Stainless steel tubes are used, with diameters down to nearly 2 mm. Some tips

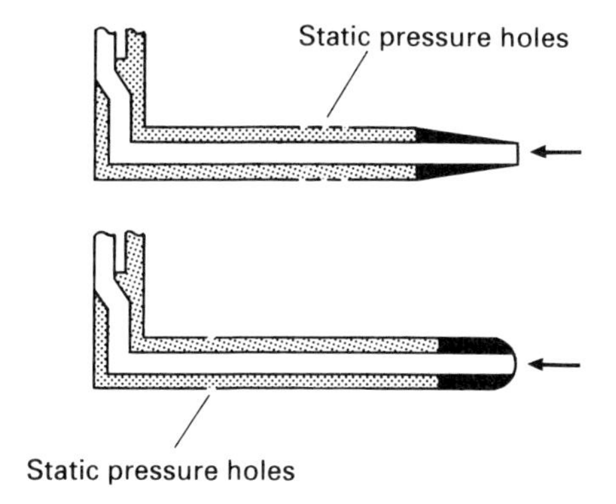

Figure 1.11 Pitot-static tube.

have the standard ellipsoidal nose which makes the probe's output relatively insensitive to errors in alignment. For velocities below about 6 m/s the differential pressure measuring device has to be a micromanometer with a resolution of about 0.25 Pa. Precision, inclined tube manometers are capable of this resolution. Above 6 m/s, portable electronic micromanometers are available.

The pitot tube's small sampling area (the orifice itself) makes it a valuable tool for investigating velocity profiles in ducts, where the portable micromanometer usually has sufficient resolution.

For more general determination of air flow there are numerous rotating vane anemometers, which use capacitance or photoelectric sensors to detect the rotation of the lightweight vanes, and thermal anemometers, which measure the cooling effect of the air on a hot-wire or thermistor element. Both can be used for measurements over about two decades of air velocity, from 0.25 m/s upwards. The rotating vane assembly provides a signal proportional to the air velocity, while the heat balance of the thermal element is a function of the square root of the air velocity, which leads to a non-linear calibration, with higher sensitivity at lower air speeds.

There are also portable versions of the direct reading, tapered tube (or variable aperture) flow meters, in which a float rises in a graduated tube of increasing diameter until the pressure loss across the aperture balances its weight. The model shown in Fig. 1.12 has a light disk moving freely on a taut axial wire, instead of the usual unrestrained, self-centring float. It is capable of an accuracy of about ±7%. Meters of this type have to be calibrated for the gas with which they are to be used, because calculation of their performance is difficult.

1.13 Acceleration sensors

Acceleration can be derived from the rate of change of velocity, of course, but it is more usual to measure this quantity by measuring the force experienced by a

Figure 1.12 Tapered-tube air-flow meter for grain beds (NIAE).

mass when subjected to acceleration, using the relationship, $F = ma$, where m is the mass, a is the acceleration and F is the resulting force in the direction of the acceleration. The mass is therefore the sensor in this case. The measurement of F is made in several ways. In some forms the mass is supported on a resilient mounting which allows it to move in one direction under the influence of the force. That direction defines the axis of the measurement. Critical damping of the assembly is necessary to avoid an oscillatory response to sharp changes in acceleration. The movement allowed may be sufficient for the displacement of the mass to be measured by a non-contacting proximity sensor, or the mounting may be strain gauged, in which case the movement is necessarily small. Alternatively, the mass can be mounted on a piezoelectric element, which generates a voltage proportional to the force generated. The piezoelectric accelerometer has the highest frequency response of the three types and the smallest size. The latter characteristic makes it the least likely to alter the dynamics of any body to which it is attached, which can be important when the purpose of the measurement is to study those dynamics.

1.14 Force and pressure sensors

Force measurement invokes either a force-balance system – long represented by two-pan scales, steelyards, spring balances and other traditional weighing devices – or measurement of pressure by hydraulic, pneumatic or electrical means. Hydraulic dynamometers and torquemeters have been established for many years in the agricultural machinery sector. Nevertheless, electrical methods of measuring force have gained ground. These generally rely on pressure measurement by the strain gauge or piezoelectric sensors introduced in sections 1.5 and 1.6.

Liquid and gas pressures (including atmosphere pressure) can also be sensed by these means. A special case of the latter is the acoustic sensor for noise measurement and analysis, which has to conform to the strict requirements of legislation on noise levels in the workplace and the environment generally (see Chapter 7).

1.15 Moisture sensors

Methods of measuring moisture in solids embrace many areas of physics and some of chemistry. In the case of crop moisture determination they also embrace numerous reference standards. The problems inherent in moisture content determination and the methods of measurement used are reviewed in Chapters 3 and 5 but here the physical background to two types of nuclear sensor sensitive to water content is outlined.

The first relates to the neutron moisture meter, which has been used by researchers and advisers for many years. This instrument employs a source of high energy neutrons (uncharged constituents of the atomic nucleus) which gradually lose their energy through scattering interactions with the nuclei of materials through which they pass. This process continues until they have lost over 99% of that energy and become *thermal neutrons*. The energy that is transferred from the neutron to the nucleus at each interaction depends on the atomic mass of the nucleus, increasing sharply as the atomic mass diminishes. Therefore the scattering effect is greatest with hydrogen – the lightest element, with a single proton as its nucleus. Since water has a high hydrogen component the ensuing thermal neutron density can be a good indication of the moisture content of the material concerned. The sensor normally used is a boron trifluoride gas filled tube, with two electrodes to which an electrical potential is applied. Thermal neutrons entering the tube react with the boron to produce α-particles which ionise the gas and create a small current impulse through the tube per α-particle. The tube therefore acts as a generator of pulses proportional to the thermal neutron flux, and these can be counted in the same way as the gamma rays mentioned in section 1.8.

The second sensor is the nuclear magnetic resonance (NMR) detector. In the 1950s NMR technique emerged as a tool for studying molecular structures in organic chemicals. By 1960 this had led to the application of low resolution NMR to the measurement of hydrogen present as water in biological materials. Since then it has been applied, inter alia, to soil water determination in the laboratory and, more recently, in the field.

The principle of NMR measurement is that an electrically charged, spinning atomic nucleus such as the proton has the properties of a magnetic dipole (i.e., a tiny magnet). When it is placed in a strong magnetic field it precesses about the direction of the field, like a spinning gyroscope, at a radio frequency dependent on the field strength and the *magnetogyric ratio* of the nucleus, which is a constant for a given element. Application of an electric field at that frequency causes resonant excitation of the spinning nucleus. On removal of the field it returns to its former precession by transfer of the acquired energy to the surrounding molecular system (spin-lattice relaxation) and by exchange with other nuclei (spin-spin relaxation) at two different rates.

In its simplest form – continuous wave NMR – a sample of the material under examination (in a non-magnetic enclosure) is placed between the poles of a large permanent magnet or electromagnet and is surrounded by a coil assembly, coupled to an RF generator. The magnetic field strength is *swept* by changing the current through an auxiliary pair of coils, coaxial with the magnet. When the correspondingly changing precession frequency of the nuclei matches the frequency of the RF generator the nuclei resonate and the amount of energy absorbed from the generator provides a measure of their abundance.

Here, as with the thermal neutron method, we are concerned with the hydrogen nucleus, although the magnetogyric ratio for fluorine, phosphorus and some other elements is not insignificant. However, the main disadvantage of the continuous wave system is that it provides no indication of the distribution of hydrogen between water in the sample and other constituents, such as oils. Neither can it take account of the effects of any paramagnetic contaminant, such as iron. This has led to the development of pulsed NMR techniques.

The single pulse measurement injects an intense pulse of RF radiation into the sample for a few microseconds. This is called the *90° pulse* because it rotates the nuclei by 90° with respect to the direction of the magnetic field. Subsequently the nuclei radiate back a complex RF signal as they return to their original state over a time period of the order of a second. The initial magnitude of that signal quantifies the total number of hydrogen nuclei in the sample. From this, and by calibration with standard samples, the percentage hydrogen per unit mass can be calculated.

The decay curve that follows the initial signal is a composite of the various spin-lattice relaxation times, but it is possible to separate these to some extent through a knowledge of the relaxation processes. For example, the relaxation for hydrogen in the solid phase of a multiphase material such as fat is much shorter than that in the liquid phase. Therefore two measurements of the return signal – one shortly after the 90° pulse and the other after the solid-phase relaxation period – can determine the distribution of hydrogen between the two phases, in principle.

From the foregoing it will be seen that NMR measurements are not confined to moisture determination. Other applications and further developments of the technique are described in Chapter 5.

1.16 Analytical sensors

pH and solute concentration

Monitoring of the acidity or alkalinity of soils and liquids of various kinds can often be done by simple colorimetric tests but when more continuous sampling and recording is required the pH meter is needed. The pH scale derives from the concentration of charged H^+ and OH^- ions present in the water. At 25°C, and expressing their concentrations in units of mol dm^{-3}, their product is a constant

$$[H^+]\,[OH^-] = 10^{-14}$$

$$\text{in pure water } [H^+] = [OH^-] = 10^{-7} \text{ and the water is neutral}$$

In acidic solutions $[H^+]$ increases and $[OH^-]$ diminishes correspondingly, while the reverse applies in alkaline solutions. Since the range of these concentrations varies enormously from strongly acidic solutions to strongly alkaline ones the pH scale was defined as the negative logarithm of the H^+ concentration, i.e.,

$$\text{pH} = -\log_{10}[H^+] \text{ or } \log_{10}(1/[H^+])$$

Thus for a neutral solution pH $= \log_{10}(10^7) = 7$ and the values for acidic and alkaline solutions are less than and greater than 7, respectively. However, the influence of temperature should be noted (Table 1.6).

Table 1.6 Temperature dependence of pH measurement.

Temperature (°C)	18	25	40	75
$[H^+][OH^-] \times 10^{14}$	0.61	1.00	2.92	16.9

The pH sensor comprises two electrode assemblies, each containing solutions of known ionic activity, in contact with wires which deliver their output to the associated meter. When these electrodes are immersed together in a solution one adopts an electric potential dependent on the pH of the solution and the other acts as a reference electrode. In combination they produce an output voltage which is a function of the pH of the solution under test, providing an attainable accuracy of ± 0.1 pH.

The calomel (mercury chloride) reference electrode is usually employed. This contains a potassium chloride (KCl) solution in contact with and saturated by calomel. Electrical connection is made by a platinum wire dipping into a pool of mercury which is in contact with the calomel. A porous plug bridges the KCl solution and the test solution. The measuring electrode employs a thin glass membrane which acquires an electric potential dependent on the pH difference between the external test solution and the electrode's internal solution. A platinum resistance thermometer sometimes completes the assembly, to provide the necessary temperature correction. Standard solutions are available for regular calibration of the sensor, which must also be regularly cleaned.

Instrument measurement of solute concentration in liquids invokes measurement of their electrical conductivity (EC). In agriculture and horticulture this is usually associated with nutrient levels, although the conductivity of milk can be an indicator of mastitis infection (see section 6.5).

The conductivity sensor is a great deal simpler than the pH sensor, since it comprises two non-reactive electrodes (commonly platinum) in a glass or metal enclosure (equally non-reactive). The assembly can be planar or cylindrical, designed as a dip cell or flow-through cell. It is made part of a bridge circuit, energised by a.c. voltage, to avoid polarisation (reverse voltage) errors that could follow from d.c. energisation, as a result of liberation of oxygen and hydrogen at the opposite electrodes. The measured resistance is the inverse of the cell's conductance (siemens) and the conductance relates to the conductivity of the solution (S/m) via the *cell constant*, K, i.e. conductivity = $K \times$ conductance. The value of K can be computed from the areas of the electrodes and their spacing and/or calibrated by filling the cell with a liquid of known conductivity such as a standard KCl solution. Once calibrated, K should not change.

Conductivity is measured in μS/m and accuracies of $\pm 2\%$ are attainable but, as with pH measurement, temperature correction is essential and cleanliness of the assembly equally vital.

The measurement is therefore fairly straightforward and it is well established industrially. On the other hand, the relationship between conductivity and the *concentration* of individual solutes is not so straightforward. In general, conductivity increases with concentration to a broad peak level, then declines again at further increases in concentration. The specific shape of the curve has to be established empirically in any particular application.

Gas analysis

Carbon dioxide concentration in air is important in some areas of greenhouse crop production and crop storage (Chapters 4 and 5). Its levels can be measured by a variety of means but for continuous monitoring and control the infra-red gas analysis (IRGA) is the established instrument. The IRGA is based on the principle that polyatomic gases and vapours absorb radiation in the infra-red region of the electromagnetic spectrum and that each has its characteristic absorption spectrum. It is used to measure the concentration of many of the gases, including ammonia, carbon monoxide, and the oxides of nitrogen, as well as water vapour, within the wavelength range 2 to 15 μm. CO_2 is measured at about 4 μm.

The sensor assembly is a tube through which the gas is drawn, which has a source of broad-band infra-red radiation at one end and a detector at the other. The wavelengths appropriate to the gas concerned are selected by an optical filter or sometimes by an interposed gas which acts in the same way. Many are double-beamed instruments, with a second gas-filled tube, containing either a non-infra-red absorbing gas as a reference or an absorbing gas for differential measurements.

The sensor's output is essentially proportional to the mole density of the gas (mol/m^3) but the meter is usually scaled in ppm (parts/million) or more precisely vpm (volumes/million) at a stated atmospheric pressure and temperature. Table 1.7 shows the minimum range and limits of detection that can be expected of commercial instruments for the gases mentioned above. A CO_2 meter should cover the range up to 3000 vpm, with an accuracy of the order of $\pm 2\%$ fro (full range output), which can be checked with standard gas samples. The response time of these instruments depends on the flow rate through the sensor tube and is commonly 10 to 20 s.

Table 1.7 Infra-red gas analysers: lower limits (ADC).

	Minimum range (ppm)	Limit of detection (ppm)
Acetylene	500	2.5
Ammonia	200	1
Carbon monoxide	30	0.2
Carbon dioxide	10	0.1
Freons	100	0.5
Methane	100	0.5
Methyl ethyl ketone	500	2.5
Nitric oxide	500	2.5
Nitrous oxide	10	0.1
Sulphur dioxide	200	1
Sulphur hexafluoride	200	1
Toluene	500	2.5
Water vapour	500	2.5

Paramagnetic oxygen analysers are used in boiler installations (Chapter 4) and crop stores (Chapter 5). Paramagnetism is the property of some materials which causes them to seek the strongest part of a magnetic field when they are in a non-uniform field. Ferromagnetic materials such as iron are extreme examples of substances with this property. Oxygen's paramagnetism was demonstrated by M. Faraday in 1851 and his apparatus was developed as a measurement tool by L. Pauling in the mid 1940s. A diagram of the Pauling *magnetodynamic* cell is shown in Fig. 1.13(a). Permanent magnets are employed to create a non-uniform magnetic field within a sampling cell. The sensor is in the form of a thin glass envelope, shaped like a dumb-bell, and filled with non-paramagnetic nitrogen. The dumb-bell is held horizontally in the magnetic field by a taut but light vertical suspension, which allows it to twist about its centre. The angle of twist can be measured with the aid of a small mirror on the suspension which reflects a beam of light, in the manner of school galvanometers. With air (or other oxygen-containing gas) in the cell the effect of the magnetic field on the oxygen causes an unbalanced force on the dumb-bell which twists the suspension until it generates an equal restoring force. In many industrial oxygen meters the well-known force-balance principle is employed. In this case the rotary movement of the dumb-bell is annulled by an equal and opposite force, generated electrically. The measurement then becomes an electrical one.

Standard industrial oxygen analysers of this type cover the range 0 to 2.5% as well as higher ranges up to 100%. Some are intended for manual control of oxygen levels but others are specifically for automatic control systems. Both types have flow meters and control valves to ensure that the sampled air (or other gas) passes through the measuring cell at the required rate. In addition, the unit for automatic control has a temperature-controlled cell for good, long-term stability. In this way, oxygen levels can be read to 0.1% concentration and the intervals for calibration checks with known gas mixtures are correspondingly longer.

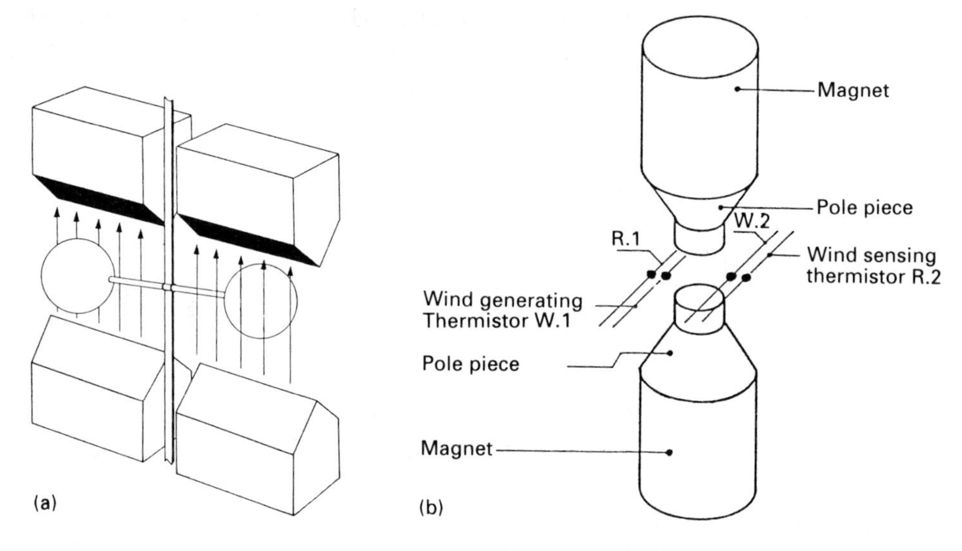

Figure 1.13 Oxygen sensors: (a) dumb-bell type; (b) thermal type (Panametrics).

A more recent development employs a *thermoparamagnetic* method, shown in Fig. 1.13(b). In this cell there are no moving parts and no tilt or flow-rate sensitivity. The sensors are four glass-coated thermistors, in pairs, located on the fringe of the strong, inhomogeneous magnetic field between the two pole pieces. The incoming gas is channelled into the region between the pole pieces, which causes the oxygen molecules in it to move towards the centre. On their inward passage they take up heat from the inner thermistors, which are slightly warmed by the current passing through them. Their increased temperature reduces their paramagnetism and they are driven from the centre by incoming molecules. That process creates a set of circulating oxygen *wind* eddies, which carry heat from the inner thermistors and warm the outer pair. The four thermistors are part of a resistance bridge network whose out-of-balance output provides a measure of the oxygen concentration of the incoming gas. Refinements of the bridge circuit reduce the effects of changes in the background (carrier) gas composition and pressure. Measurement ranges down to 0–1% O_2 are available. In association with microcomputer software for signal processing the sensor is capable of an accuracy of ±2% fro in that range, and ±0.2% fro for concentrations over 90%. Over the intervening spans it can achieve ±1% fro. Flow rates between 0.05 and 1 l/min are accommodated and the time constant is less than 5 s. A basic version of this sensor has an accuracy of ±2% fro over its central ranges and a time constant less than 40 s.

Another method of oxygen analysis is widely used to measure the emissions from boilers and automobile engines. This employs a ceramic sensor based on zirconium oxide, which acts as a solid electrolyte at temperatures around 800°C, i.e. it conducts oxygen ions at that temperature. When the ceramic element is sandwiched between two thin platinum electrodes it produces a voltage which is proportional to the logarithmic difference in oxygen concentration between the exhaust gas and a reference, usually atmospheric air.

Near infra-red transmission and reflectance

The near infra-red (NIR) waveband (1 to 2.5 μm approximately) provides the basis of an increasing number of tools for quality analysis in agriculture and food processing, arising from pioneering work by K.H. Norris (US Department of Agriculture) in the 1960s. The main technique employed is to illuminate a sample of the material to be analysed with infra-red radiation, and to measure the proportion of that radiation reflected at a series of wavelengths, selected by a spectrum analyser. The selection of these wavelengths for a particular application is based on experiment and much ingenuity has been employed to identify the wavelengths which most readily provide an acceptable correlation with the results of chemical or physical analysis. In some cases the process has been reduced to comparison of reflectance at two wavelengths only, for a particular constituent of the material. Computer programs have been developed to speed up the search for suitable pairs of wavelengths. As a result of this activity, over the past 30 years routines for measurement of protein, fat, moisture in plant material and soils, forage and silage constituents have been established. Detection of grain infestation has been demonstrated, too.

Near infra-red transmission techniques have been developed in parallel with reflectance. Grain and oil seed moisture, protein and oil content can be measured in this way. The material has to be in a thin layer in this case, since its transmission is low.

Ion-selective sensors

Developing semiconductor technology gave rise to a form of transistor known as the field effect transistor (FET), which found particular application in circuits where a high input impedance was advantageous. That in turn gave rise to the ion-selective FET, or ISFET, as a candidate for replacing the established ion-selective electrochemical cells, including the pH cell and others for measurement of chloride, nitrates, potassium and sodium, inter alia.

The FET, like other transistors, has three electrodes, one of which controls the current between the other two. In the FET the controlling electrode is known as the *gate* and in the ISFET the gate becomes a sensor, through the formation of a coating on it which reacts electrochemically with a specific ion when the two make contact. The effect on the FET is to vary its current according to the concentration of the sensed ion.

Ion-selective field effect transistors have advantages of small size and freedom from the liquids and fragile glassware of the electrochemical cells. They can also be made in arrays capable of responding to a variety of stimuli – a feature that is being exploited in the development of *electronic noses* for many purposes, including pollution monitoring (see Chapters 4 and 7).

1.17 Image analysis

Optical analysis has been applied for many years to measurement of droplet distributions, leaf areas, particle motion and many other quantities of interest to agriculture and its related industries, but the techniques have been mainly reserved for *off-line* measurements, particularly in the laboratory or test centre. Following the development of the video camera a new and very versatile sensor became available to these industries and it is already the basis for the increasing number of on-line image analysis systems discussed in later chapters. The solid-state image sensor has largely taken over from the imaging tube of earlier cameras, by virtue of its smaller size, lower cost and greater robustness. The sensor is a flat semiconductor element in which has been formed a matrix of separate light-sensitive diodes, commonly with surface dimensions of about 10 μm × 10 μm. After conversion from analogue to digital form their output is read at intervals to provide the data for a succession of television-type frames which are subsequently processed by a computer before being passed on to a display monitor or a control system. The charge coupled device (CCD) camera is the predominant form of the solid state camera. Scanning of the *pixels* is accomplished by sequential transfer of electrical charges representing the light levels at the pixel sites.

The pixel diodes are capable of responding to a wide range of light levels but their output does not automatically incorporate colour information. However, they can be coated with optical filter material to provide interleaved matrices which provide a conventional red-green-blue (RGB) output for colour monitors.

The generalised form of an image capture and analysis system is shown in Fig. 1.14. The output from the camera passes to a *frame grabber* containing the master clock which synchronises the sampling of data from the sensor and its assembly

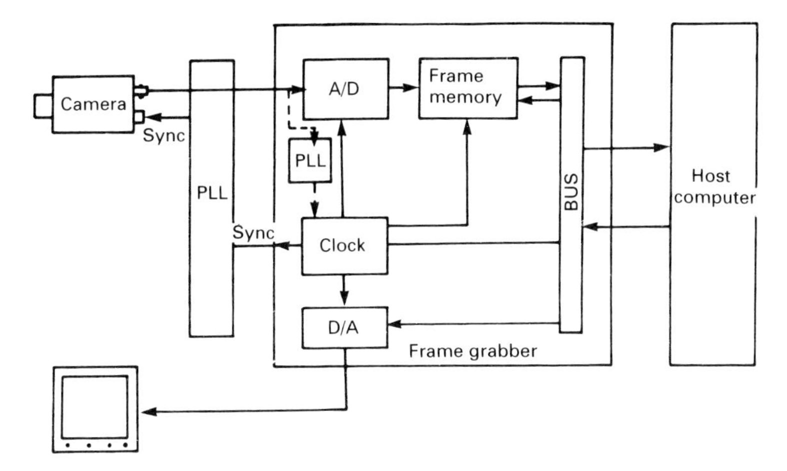

Figure 1.14 Image capture and analysis system (Photonics Spectra).

into sequential frames via a phase lock loop (PLL). The output from the sensor is digitised before it is passed into the frame memory. Each frame of data then transfers to the computer via the bus (data highway). After processing in the computer each frame is then reconverted to analogue form in order to make it compatible with the standard TV monitor.

1.18 Further information

Standard college textbooks and course notes will provide the necessary information for those who wish to go more deeply into the mathematical and physical background to sensors and measurement theory. Up-to-date publishers' lists are increasingly available through their home pages on the Internet.

Developments in measurement standards and calibrations can be followed through contact with national and international standards organisations. In the UK, the primary source is Information Services at the NPL, Teddington, Middlesex, TW11 0LW, which can supply a very useful range of booklets on standards of measurement and the measurement/calibration services that it offers. The 68-page *Guide to the Measurement of Humidity*, published by the Institute of Measurement & Control (18 Gower Street, London WC1 6AA) and prepared by NPL, is particularly recommended.

Information on developments of new sensing and measurement techniques can only be acquired through wide reading of scientific and technical journals. Many of the techniques and devices currently employed in agriculture and related industries have been imported from the wider sphere of industrial instrumentation. Development of specialist equipment for a relatively small and widely dispersed market such as agriculture is never easy and it is more likely that advances will continue to come by adoption, particularly from the chemical and biochemical industries. Developments in these areas, coupled with optoelectronics, are creating new opportunities for monitoring the complex materials and processes encountered in plant and animal production, in association with the advanced data processing techniques outlined in the next chapter.

Exploration of this wide field has been made much more practical by the development of the Internet's resources and the means of accessing them. *Nature* (http://www.nature.com) and *New Scientist* (http//www.newscientist.com) are two well known science journals accessible in this way. The UK Institute of Physics (http//www.iop.org) provides access to three of its general journals, *Physics World, Opto & Laser Europe* and *Scientific Computing*, all of which cover new developments in sensors and their applications. Web browsers, employing hyperlinks, provide convenient entrée to many other servers, worldwide.

Chapter 2
Instrumentation and Control Systems

2.1 Computer-based instrumentation

The general form of a contemporary instrumentation system is shown in Fig. 2.1. This brings together the successive steps from the data derived from the sensors through to the display and printout of the processed data, via data links of several kinds. These steps will be taken in turn.

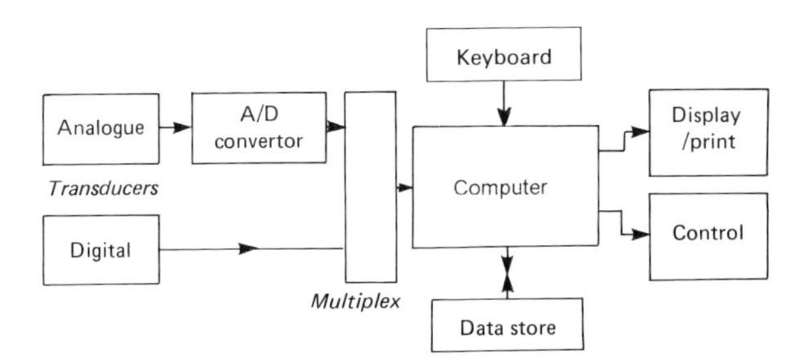

Figure 2.1 Elements of a computer-based monitoring and control system.

The transducers

This embraces sensors and any electrical/electronic conditioning of their output that takes place in the transducer unit. For example, digital outputs may be regulated to a specific pulse width and height, while analogue outputs may be amplified to provide a 0–10 V d.c. output or the noise-immune 4–20 mA current output that is a common industrial format for transmission between transducers and remote meters.

The analogue/digital converter (ADC)

The ADC is now a mass-produced, inexpensive microcircuit device, available in many forms – as is its opposite number, the digital/analogue converter (DAC). The ADC commonly converts the sampled analogue signal into between 10 and 24 bits. For comparison, a 12-bit number covers the decimal range 0–4095, so the least significant bit (LSB) represents better than 1 part in 4000 of the full range of the device. Higher resolutions are attainable with the larger bit – numbers, of course. That can be important if small differentials in a large signal are of practical interest. Some ADCs are linear to the LSB, others guarantee linearity to better than 0.05%. Most operate from input voltages up to about 5 V but some have built-in amplifiers. Their speed of operation varies widely, too, from 100 sps (samples/second) to over 1 Msps. Their outputs are usually in serial pulse form but some also have parallel data outputs for faster data transfer. The speed of conversion determines the sampling rate and hence the highest frequency component in the transducer output that can be reproduced. However, even a high sampling rate may not avoid the problem of aliasing, depicted in Fig. 2.2. Here the sampling of the signal creates a spurious low-frequency component from a high-frequency component which is otherwise of no significance in the measurement context. This source of error can be removed by an anti-aliasing filter, incorporated ahead of the data analysis stage.

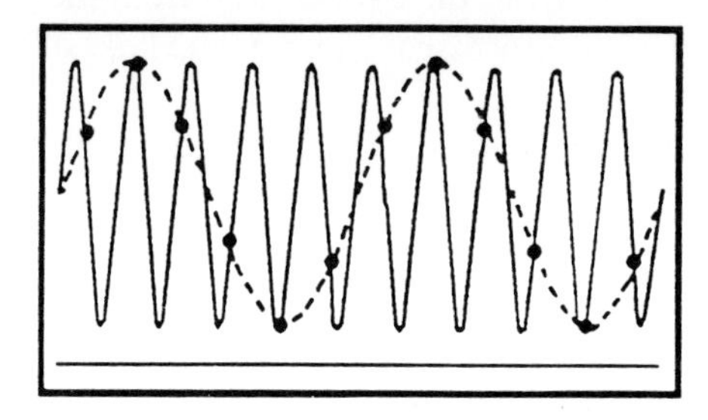

Figure 2.2 Aliasing.

Manufacturers also warn that the ADC is particularly prone to the existence of *ground loops* in the system. The ground or (earth) loop is a common problem in instrument systems, especially those that require high immunity from pick up of unwanted signals. It can occur when several parts of the system are grounded at different points, with the result that currents can flow between those points towards some common *sink*. In consequence, the ground current from one part of the system can affect the ground level potential of another part, thereby creating spurious signals in the latter section. In the case of ADCs the effect is to downgrade their resolution, since the reliability of their least significant bits is impaired. The recommended course in this context is to have separate analogue and digital grounds and to connect them at only one point.

The multiplexer (MUX)

The MUX is a set of semiconductor switches which are turned on and off by a circuit driven by clock pulses, to deliver the transducer inputs serially to its output. It can be upstream of the ADC if all the transducer outputs are of the analogue type and, indeed, it is incorporated in some ADCs. In general it may be required to transfer signals at millivolt levels, or even microvolt levels, therefore it must create negligible distortion and noise. It may also have to operate at MHz clock frequencies.

Like the ADC, the MUX may incorporate a third common instrumentation module, the operational amplifier (op amp, for short) which serves as more than a signal amplifier. It is often used to isolate (buffer) succeeding stages of a system from *common mode* signals – i.e., potentials picked up equally at the input terminals – by virtue of its differential input, which ensures that they go no further. Therefore its common mode rejection ratio (CMRR) can be an important aspect of its performance. Equally – and for the same reason – it can isolate the output of the preceding transducer from ground potential, as required in some measurement systems. In the latter capacity it can be invaluable for bridge and thermocouple measurements. It is a component in many of the instrumentation and control systems mentioned in this book, although it only appears overtly in Fig. 3.5(a), where it is represented in the standard way as a triangle with two inputs and one output.

The computer

Any specific details about the computer are almost certain to be out of date in a very short time, but it is possible to venture some comments on the direction in which computer-based instrumentation systems are going. It is noteworthy that portable computers now have sufficient processing speed and back-up memory in one form or another to suggest that the *virtual instruments* now available will become commonplace in the near future. The instantly recognisable feature of these instruments is their mimic control panels and mimic meters on the computer's monitor (Fig. 2.3). Indeed, this feature is already widely available in multimedia computer systems, for selection of CD ROM tracks and adjustment of control settings, by keyboard or *mouse*. These are examples of the graphical user interface (GUI) which is now a feature of many applications of computers.

The software elements of a data acquisition system (DAQ) with the virtual instrument panel are an editor for assembling the elements of the panel and another for connecting them so that they interact with the data sources and carry out the processing functions, as required. Most of these systems operate in the Microsoft® Windows™ environment. They receive multiplexed data from interface boards or cards via serial or parallel communications links, using the host computer's standard serial and parallel ports, such as those used for fax/modems and printers. Many use the Windows95™ configuration standard for plug and play (sometimes abbreviated to PnP, for Plug'n Play), which greatly simplifies these connections. At start up a PnP board transmits its required resources to the host computer, which automatically allocates them to the board. Another common feature is support for dynamic data exchange (DDE), which allows data to be swapped between applications.

The Personal Computer Memory Card International Association (PCMCIA) standard cards are probably the most widely employed interface elements, by virtue of their range of applications as input/output devices to computers from notebook and lap-top size upwards. A single card might support 16 single-ended

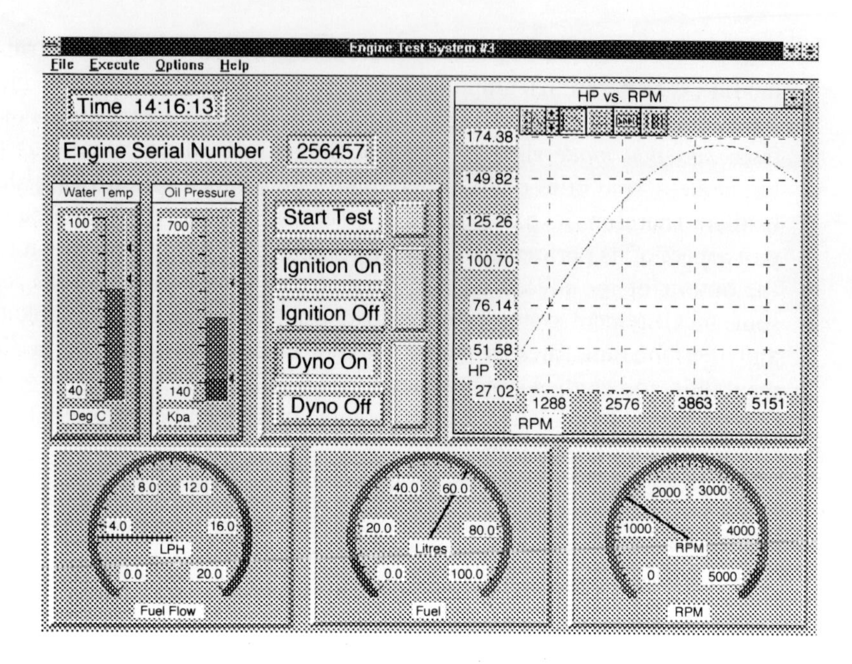

Figure 2.3 Virtual instrument (Intelligent Instrumentation).

or 8 differential inputs or outputs, together with a time/counter and a triggering function. More elaborate boards may contain a local digital signal processor (DSP) which performs preliminary processing such as filtering or spectral analysis of the signals.

When a board is connected to a PC with the fast peripheral component interface (PCI) bus the frame grabber required for an image processing application does not need the local memory shown in Fig. 1.14. It can simply digitise the camera's video signal in real time and pass the data directly to the host computer's main memory. The PCI standard was first developed in 1991, with wide industry support and it was designed *ab initio* to provide plug and play capability in full-size computer boards in the way that the PCMCIA did for the plug-in card.

2.2 Data transfer

The PCI bus is one of many formal or *de facto* standards for interfacing and information transfer between computers and peripheral devices, or between computers. They are concerned to differing degrees with the physical means by which data are transmitted and received, the way in which data are handled and the protocols. Protocols govern starting and stopping of data transmission, error detection and correction, identification of talkers (senders) and listeners (receivers) and the generation of special functions such as requests for device status or acknowledgements. Some of these standards have been initiated by specific vendors; others have been the result of cooperation between suppliers, either alone or acting with users and standardising organisations, up to and including the International Organisation for Standardisation (ISO). Substantial progress has been achieved in establishing standards that are not vendor-specific, but the rate of change in computer technology has required modification of existing standards and the introduction of new draft standards to keep abreast of these changes. Upgrades face the usual problem that they have to be both backward and forward compatible as far as possible, in order to support installed equipment operating to the earlier standard, while new standards have to coexist with older ones over a period of years.

ISA and PCI

At the individual computer level the main external interfaces are the familiar serial and parallel connectors that are used to couple in the mouse, the printer, the fax/modem and other common peripherals. These appear to serve their purpose well enough, but at the internal board level the internal systems adapter (ISA) bus which appeared in IBM computers in the early 1980s is giving way to the PCI bus. The former provides 16-bit parallel data transfer and 8 MHz clock speed, so that it can support transfers at 16 Mb/s (b = byte and 1 byte = 8 bits, or 0s and 1s). However, with increasing emphasis on graphics images requiring up to 1 Mb of data apiece, the ISA bus slows down the system. By contrast, the PCI bus has introduced clock rates of 33 MHz, with data transfer at 132 Mb/s in the 32-bit (4-byte) version. A 64-bit version (264 Mb/s at 33 MHz) extends the opportunities to process full-motion images in real time, as well as enabling data acquisition from signals in the radio frequency band. Beyond this, the PCI standard was designed to resist obsolescence.

Nevertheless, there are many circumstances in which such high rates of data transfer are not required and other standards are more appropriate. Among these, two of the most commonly employed are introduced here.

IEEE-488 and RS232

Many programmable laboratory and industrial instruments from many suppliers employ versions of the IEEE-488 interface, standardised by the USA's Institute of Electrical and Electronic Engineers in 1975, and also known as the general purpose interface bus (GPIB), which it was designed to be. It was later adopted by the International Electrotechnical Commission as IEC standard 625. It employs 16 identical lines, divided functionally into three groups. An 8-line, input/output data bus carries the data bytes being transferred, together with any interface control messages. This information is in bit-parallel, byte-serial form – i.e., the 8 bits of each byte are transferred simultaneously on the eight lines, to be followed by the eight bits in the next byte. A three-line transfer control bus operates the *handshake* process which ensures that a specific sequence of signals is exchanged between the talker and the listener before, during and after the data transfer. The remaining five lines constitute a management bus which distinguishes between data, bytes and command bytes, and which manages the traffic between linked instruments on the bus. Data transfer rates up to 1 Mb/s are attainable.

The original IEEE-488 simplified and standardised the interconnection of instrumentation at the electrical, mechanical and protocol levels of a GPIB. In the late 1980s IEEE-488.2 added two further layers of standardisation, to avoid different approaches by different suppliers. One of these defined standard data codes and formats, the other established a set of commands and queries for all instruments and controller requirements. Subsequently a consortium of major suppliers used IEEE-488.2 as a basis for a single, comprehensive command set for all programmable instruments. More recently, application specific integrated circuit (ASIC) chips have been developed which implement a high-speed IEEE-488 to increase its transfer rate to 8 Mb/s.

The GPIB is generally employed for short-range communications. For longer distances the USA's Electronic Industries Association (EIA) established several serial links which have become established standards. The best known and probably the most widely employed is RS232, introduced in 1969, which is the serial port (interface) built into most computers. It employs only two signal lines connecting the talker and receiver but the complete port has nine lines, as shown in Table 2.1. Its theoretical maximum length is 30 m and its maximum transmission rate 2.4 kb/s, but much longer lengths are possible at lower transmission rates. In electrically noisy environments an externally screened cable is essential. The screen is preferably grounded at one end only (the listener end in a one-way data link).

As with the first version of IEEE-488 the RS232 standard specifies only the physical aspects of the interface and the signal characteristics for the binary states 0 and 1 (a positive and negative voltage between 5 and 15 V, respectively). It also

Table 2.1 RS232C signal definitions.

Pin no	Name	Function
1	FG	Frame ground
2	TD	Transmitted data
3	RD	Received data
4	RTS	Request to send
5	CTS	Clear to send
6	DSR	Data set ready
7	SG	Signal ground
8	DCD	Data carrier detect
20	DTR	Data terminal ready

operates in the asynchronous mode – meaning that the talker and listener are not synchronised by timing pulses from a common digital clock: the transmitted sequence of bits carries the necessary timing information. The most common way of doing this is to transmit the data in ISO's 7-bit (plus parity bit) International Data code, which is derived from the American Standard Code for Information Exchange (ASCII code) produced by the American National Standards Institute (ANSI). Each 7-bit data word is framed by a *start* bit (high level) and one or two *stop* bits (low level). The word is accompanied by the parity bit, which provides a means for error detection. It is automatically inserted by the sender to bring each byte to odd or even parity, i.e., to make the number of 1s in the byte odd or even, respectively. Any word that fails the listener's parity check is immediately diagnosed as corrupted and an error-correcting procedure is initiated. The latter involves retransmission by the sender. This procedure will not detect all errors because it is possible for two or more bits in a data byte to be corrupted and, by chance, to produce the required parity. More elaborate checks on blocks of data can overcome that limitation but these checks require longer transmission times, [*Note* The ISO code employs even parity. Seven-bit ASCII is now commonly referred to as plain ASCII in the context of general information transfer.]

RS422 and 485

Cable runs of up to 1500 m are possible with two other well known EIA serial links, RS422 (related to RS232C but with the polarities of the logic levels reversed) and RS485 (for use in multipoint systems). These employ twisted-pair cabling, supplying a balanced signal to the differential inputs of op-amps at their interfaces. The twisted-pair conformation improves the noise immunity of the link over these lengths and the op-amp's CMRR helps to isolate the system from common-mode interference to some extent. Nevertheless, ground potential differences can be high between the extremities of the cable and even the op-amp's differential input can be insufficient to protect the system against damaging noise voltages or electrostatic pick-up. Linking device grounds can reduce common-mode noise but this can create another problem – the resulting ground loop can radiate sufficient RF noise to breach EMC limits (see later in this chapter).

Data transfer rates can be much higher than those recommended for RS232 (up to Mb/s) but at higher rates the high-frequency characteristics of the cable become increasingly important. Whereas at low frequencies the electrical resistance of the cable is its limiting feature, since it determines the voltage reduction along its length at a given current, at high frequencies its impedance (a function of its resistance, inductance and capacitance) must also be considered. The cable behaves as a transmission line which has to be terminated by a resistance close to its characteristic impedance, normally of the order of 100 Ω. Without that termination an impedance mismatch would occur at the receiving end and reflection of the signal would result, leading to interference with the incoming data stream.

Termination becomes more complex when the system is a multipoint one, with several driver/receivers coupled to the same line. Each lateral stub represents a separate transmission line. Ideally, therefore, stub lengths should be as short as possible.

Networks

Many industrial process management and control systems rely heavily on digital networks to link supervisory control and data analysis (SCADA) functions extending from the level of overall management to the shop floor. Standardisation of the hardware and software aspects of these networks has been an ongoing process since the early 1980s, with the goal of making them increasingly vendor-independent or *open*. The ISO's Open Systems Interconnection (OSI) emerging standard has provided the guidelines and target for much of this development.

Open Systems Interconnection is based on a seven-layer model, of which the lowest level (1) specifies the physical layer (cables and connectors) and the second layer (level 2) deals with the data link's format and line control. Higher levels cover transmission protocols, which embrace TCP/IP (Transport Control Protocol/Internet Protocol), both of which are involved in communications on the Internet. The uppermost level (7) is the applications layer, which is the subject of the last part of this subsection.

Ethernet

The industry requirement for local area networks (LANs) led, inter alia, to the development of Ethernet, which was introduced by Xerox, DEC and Intel and which became the basis for another IEEE standard, published in the mid 1980s. IEEE-802.3 specified two versions. The full (or thick) Ethernet employs twin-sheathed cable while the thin (or Cheapernet) version employs less costly cable as well as less costly connectors at the *nodes* – i.e. at the tapping points to the individual stations served by the network. Both operate at 10 Mb/s data transfer rate but the thin version has a shorter segment length (200 m against 500 m) and supports fewer nodes per segment.

Figure 2.4 is a block diagram of an Ethernet node showing the transmit and receive stages between the LAN and a local or higher level bus. The collision detector is necessary because Ethernet normally operates in the multiple access mode, in which a transmitted message goes to all the other nodes in the network. To avoid all of the talkers talking at the same time, a station with a packet of data to transmit first checks whether another station is transmitting, using a technique called carrier sensing. If it detects that condition it defers its transmission until the bus is free. However, that cannot prevent two stations starting to transmit almost simultaneously, thereby creating a collision situation. When this happens the collision detector causes the stations to delay their transmissions by different amounts and single-file queuing ensues.

An Ethernet-based system (IEEE-802.4) has also been the basis of the

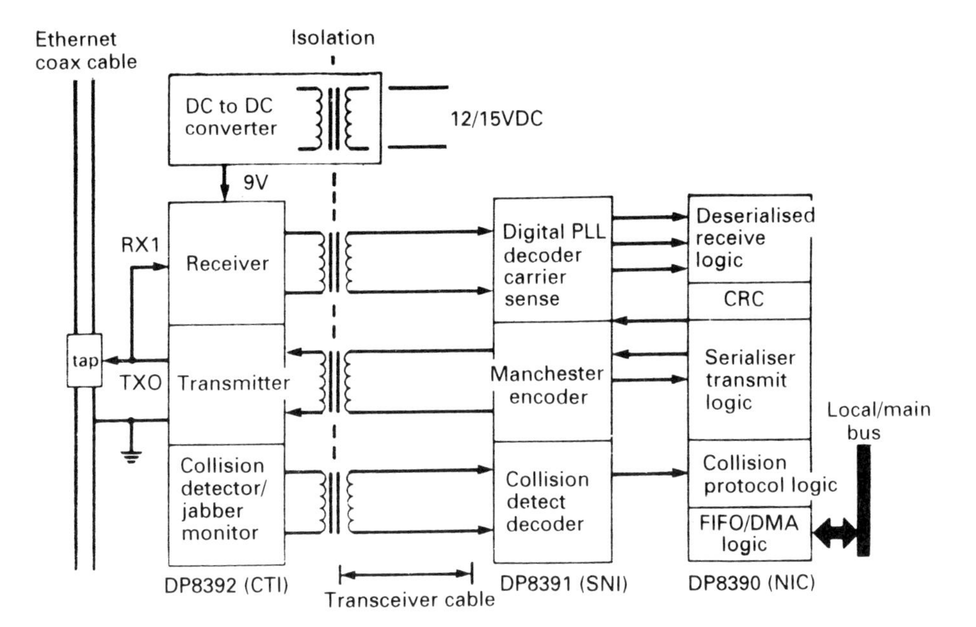

Figure 2.4 Ethernet node.

manufacturing automation control (MAC) system introduced by General Motors in the 1980s for automated processes in manufacturing industry. This employs token passing, which ensures that stations transmit data in a predetermined order. Only the station with the *token* at a particular time can transmit a message, so there is no danger of collisions.

Factory token-passing networks were operating at transmission rates of 10 Mb/s and higher in the early 1980s. However, the fast carrier sensing, multiple access/collision detector (CSMA/CD) Ethernet of the 1990s operates at the same speed and even faster versions are coming forward. Switched LANs make improved use of the bus through the incorporation of an electronic switch which intercepts all messages on the bus, determines their destination and forwards each one only to the required address. This greatly reduces the amount of blocking that occurs relative to the CSMA/CD system.

Fieldbus and Profibus

A full SCADA system for a multisensor, multiactuator industrial process will include all of the elements shown in Fig. 2.5. Levels 1 and 2 constitute a networked client/server configuration in which the server responds to requests from the clients for data on the state of the process. Level 1 clients will be concerned with individual aspects of management decision making, including process control strategies. Level 3 contains the field devices (sensors and actuators) and their on-line, real time monitoring and control units (manual and automatic) which also provide the data for the higher levels.

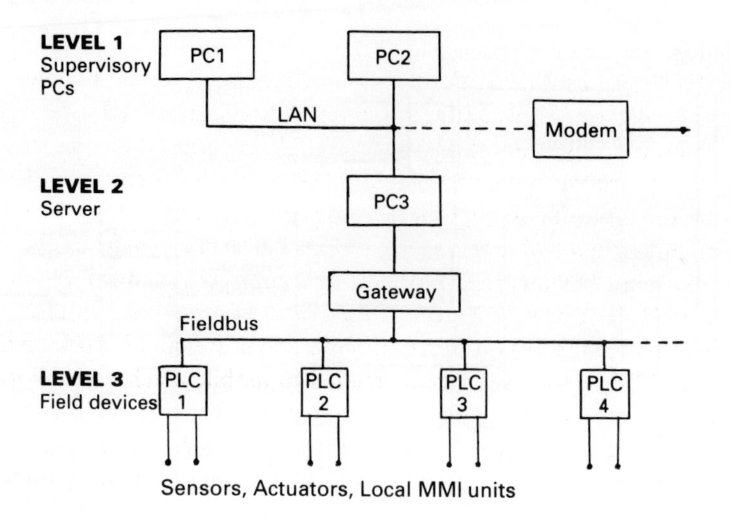

Figure 2.5 SCADA system.

Levels 2 and 3 are shown connected by a *gateway*, which is a computer unit that mediates any change in protocol required between two levels of the system. A gateway is sometimes required when there are subsystems in level 3.

The requirement for a serial, digital link of lower cost and power requirement than MAP and Ethernet at level 3 led to the search for a suitable international standard. In 1985 the IEC produced a draft standard for what has become known as the Fieldbus, operating at two speeds – 31.25 kbits/s and 1 Mbits/s. [*Note* bits/s (or bauds), not bytes/s.] In the following year the Instrument Society of America (ISA) reactivated its SP50 Committee, which had produced the 4–20 mA analogue current standard, referred to at the start of this chapter. This group of manufacturers and users produced a set of requirements similar to those of the IEC and they evaluated several Fieldbus proposals. The latter were based on layers 1, 2 and 7 of the ISO/OSI model.

By 1990 a draft proposal for the physical layer (layer 1) was available for comment but consensus on the other aspects had not been reached. Meanwhile, from 1987 onwards the initiative of a European user group led to the development of a similar concept, known as Profibus (*Pro*cess *fi*eld *bus*) which has obtained *de facto* standard status in Europe, the USA and elsewhere. Profibus supports multivendor systems with network components supplied by many major manufacturers of field level devices. However, there is continuing support for the IEC/ISA Fieldbus. Its physical layer has been tested for reliability in the presence of electrical noise; freedom from *crosstalk* between separate signals and the ability to cope with transmission errors generally.

Overall, it is clear that a single practical international standard to suit all users in all circumstances is hard to define. Nevertheless, a considerable degree of *openness* has already been obtained through industry cooperation.

Controller area networks

The controller area network (CAN) was developed by Bosch in the mid 1980s and was adopted as a German DIN Standard. At about the same time the Society of Automotive Engineers (SAE) in the USA introduced Recommended Practice J1708 for serial data communications (1986). Both systems were designed for the automotive world, to provide a reliable, serial communications link between autonomous control modules in an electrically noisy environment. By extension they have an obvious role in tractor and implement control.

In 1991 the ISO's Technical Committee 23 (agriculture and forestry) set up a subcommittee (SC19) on electronics in agriculture, and a working group (WG1) was established to cover aspects related to mobile machinery. Working group 1 first produced a standard for electrical plugs and sockets as layer 1 of a projected 11-layer model. The group then moved towards standardisation of a bus system based on the CAN, version 2.0. This shares much common ground with SAE's recommendation for trucks and buses (vehicular), J1939, which is also based on CAN 2.0. More details of these developments are given in Chapter 3.

Long-range transmission

The full SCADA system shown in Fig. 2.5 includes a link to a remote terminal via the telephone network. This is an increasingly common way of collecting meteorological and other field data, either directly from a data logger or from an off-line portable computer, via a PCMCIA card connector. Equally, the line is sometimes used to download data to a remote site, or even to provide periodic supervisory control of on-line equipment at the site.

The modem

This device provides the necessary communicating link between transmitters and receivers of digital data and the telephone network, which is primarily designed for audio-frequency traffic in analogue form. Its primary function is to *mo*dulate and *dem*odulate an audio-frequency signal (hence *modem*) which carries the digital data in analogue form along the line from the transmitter to the receiver. Outgoing data are modulated and incoming data are demodulated.

Modem performance has been standardised by the Comité Consultatif International Téléphonique (CCITT) – a role now overseen by the International Telegraphic Union (ITU). Table 2.2 lists some of these standards and their data transfer rates in bits/s or bauds. [*Note* *bis* is twice in French.] It is immediately noticeable that data transfer rates are low, compared with previously mentioned methods of serial transmission. This is a consequence of the inevitable level of electrical noise on the telephone lines, which sets an upper limit of about 35 kbaud. It follows that the transfer of 1 Mbyte (8 Mbits) of data between two V32 bis (14 400 baud) modems will take about 10 minutes, while a V22 bis (2400 baud) modem at one end of the line would increase that time to about an hour, even if a V32 bis unit was at the other end.

Table 2.2 Modem data rates.

ITU/CCITT standard	Baud rate (data)
V22	1200
V22 bis	2400
V32	9600
V32 bis	14400
V34	28800*
V42	33600*

* Data compression techniques can quadruple these rates.

Substantial reductions in transmission times can be achieved by data compression, which is a technique widely used to speed up transfer of large data files over networks. This technique requires the sending device to recognise sequences of identical characters and to replace them by two characters – one representing the character itself and the other the number of times it occurs in a sequence. For example, the decimal equivalent of a space in ASCII/ISO 7-bit code is 32 (0100 000 in binary form), therefore ten spaces would be compressed to 10 32. Transmission rates can be increased severalfold in this way, depending on the opportunities for compression.

Figure 2.6 illustrates the functions of a modem, connecting the serial outlet (RS232C) of a computer to a telephone line. Outgoing data from the computer are compressed, then divided into frames or blocks before passing to the modulator. Incoming data are demodulated and decompressed. The remaining functions are concerned with error detection through the checksum procedure.

Parity checking of 7-bit ASCII coded bytes was introduced earlier in this section, in relation to RS232C. There it was noted that if noise corrupts more than one data bit in a byte parity checks can fail to detect the error. Normally, noise affects more than one byte in a data block and it only requires one detected error in a block to initiate its entire retransmission. Therefore block by block checking reduces the chance that cancellation of error in bytes will disguise transmission errors. Nonetheless, any corrupted data can cause major problems. In addition, not all data are transmitted in plain ASCII code: many forms of binary file use all eight bits in a byte for data, leaving no room for a parity bit. The checksum created by the modem during transmission overcomes some of the limitations of the parity check by adding a measure of the total bytes in a block to the end of the block. The receiving modem computes a checksum on the demodulated data received and if the two sums do not agree it calls for retransmission of the block from the buffer store.

Even this check does not eliminate the possibility of undetected errors due to chance cancellation. As a result, many modems now employ an even more rigorous check, the cyclic redundancy check (CRC) which is dealt with in Chapter 3.

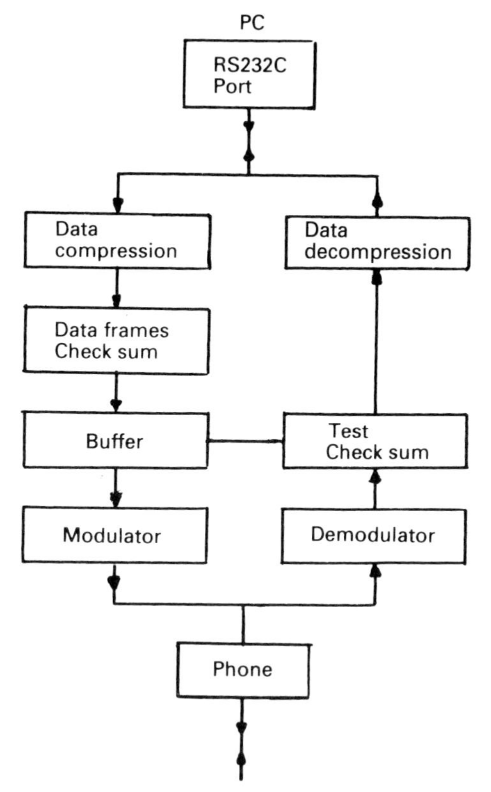

Figure 2.6 Modem functions.

Software associated with modems comprises the file transfer protocols and the command set, which must be common to the transmitter and receiver. The best known of the former are the Xmodem and Zmodem protocols. The Xmodem has largely graduated from checksum to CRC error checking. The Zmodem protocol is an improved Xmodem with faster throughput and recovery of corrupted transmissions. The AT command set developed by Hayes in the USA is the *de facto* standard, i.e. most modems are Hayes-compatible. The commands control dialling and hang up, selection of tone or pulse dialling, error correction and data compression, error messages and much else. All commands start with AT (short for *at*tention).

ISDN

In the 1980s the CCITT established the framework of a world-wide digital communications network for simultaneous transmission of voice, data, text, fax and video information. The resulting Integrated Services Digital Network (ISDN) is now an established service, commonly used for multimedia applications such as video conferencing. The European Telecommunications Standards Institute (ETSI) has established EURO-ISDN, which has two independent 64 kbaud channels and a 16 kbaud control channel, time shared to operate at a total 144 kbaud. Transmission of files by this route can be over three times faster than via an analogue modem. A PC is coupled to an ISDN line through a special interface.

Wireless transmission

Direct ground-to-ground radio telemetry of information from the field to a central monitoring and control site has a long history in agriculture, using techniques such as frequency multiplexing to convey several channels of analogue and/or digital data almost simultaneously. Equipment for this purpose has moved into the UHF region (450 MHz) and two-way links operating at 9.6 kbaud are available for point-to-point data transmission over several miles. More recently, indirect ground-to-ground communication via satellite has become more economic, in the same way as the satellite navigational systems now widely used on land or at sea. With the aid of portable transmitters/receivers messages can be exchanged between mobile sites and remote equipment can be monitored.

Low-cost LANs, operating on radio or infra-red links, are covered by a 1997 industry standard, IEEE-802.11. An access controller links to each PC via a PCMCIA card or ISA bus.

Interfacing standards

Preceding parts of this section have outlined the ways in which progress has been made towards global, *open* systems of communications and data transfer in measurement and process control technology. However, in any system there are interfaces between devices such as measuring instruments, recorders, printers and displays, which require specific *drivers*.

Many drivers are vendor-dependent and their numbers increase as vendors update their equipment. This creates a major barrier to *openness* and it has caused much duplication of effort in developing interfaces and drivers to link applications and devices from different vendors. To address the problem, in 1995 Microsoft and five process control vendors produced a draft standard for the industry aimed at elimination of this duplication. It is based on Microsoft's object linking and embedding (OLE) software. This is commonly used for importing copies of *objects* such as files, spreadsheets and pictures from one application into another, given that both support OLE. Subsequent changes made to an object in the source application can automatically update the copy in the recipient application.

Although OLE has greatly facilitated software development and integration it was recognised that it needed extension to make it compatible with the requirements of process control. The draft standard, termed OPC (OLE for process control), set out to provide a common interfacing standard between the servers of Fig. 2.5 and OLE-compatible applications at client level. The longer-term agenda is to seek support from existing standards organisations or to form an independent, non-profit-making organisation to promote the standard, which has been described as the final step to open systems.

2.3 Control systems

In this section control systems are divided, somewhat arbitrarily, into those for sequential control of a process and those which exert effectively continuous regulation on the process. The former are characterised by a sequence of time intervals – each triggered and terminated by an event occurring during the process – within which the process advances another step. The latter employ feedback from the process to adjust its inputs in such a way that the output achieves an acceptable approximation to the required level or time profile. [*Note* As in the measurement process, no control system can achieve absolute accuracy, except in the straightforward handling of unit items.] Many process control operations involve both types of control, working interactively.

Programmable logic controller (PLC)

Many sequential control systems are still based on electromagnetic relays, switched by inputs from manual and automatic starters, timers and other event sensors, to control output devices such as motors, heaters, fans and batching devices. These *hard-wired* systems can be built from mass-produced, quality-controlled components to provide low-cost systems, reliable even in adverse environmental conditions, such as those found in some agricultural environments. However, changes to the system can involve major rewiring and retesting to ensure that the required control logic (sequence of *ons* and *offs*) has been achieved. The comparative simplicity of making changes by software programming led to the introduction of the PLC in the early 1970s, since when it has become a major component in industrial and domestic equipment of many types. Over the intervening years the increasing market, combined with developments in electronics, has made the PLC ever smaller, cheaper, faster and more powerful in many ways. Nevertheless its basic format (Fig. 2.7) has not changed markedly.

Hardware

The main unit (PC110 in Fig. 2.7) contains a microcomputer with the usual built-in digital clock which controls the successive program steps made by the computer's central processing unit (CPU) and, after frequency division, provides a clock pulse input to any process timers that are built into the system. This unit also has the memory required for the program instructions and for data on the logic states of the system's inputs and outputs. The PC110 has a set of terminals for connection to input and output devices, which can be increased by coupled expanders (PCE111 in Fig. 2.7) to a given limit. The computer is programmed via a keyboard/display unit (NLPL180 program loader) which is either plugged directly into the computer or coupled by cable via the adapter (NLLA185). The loader also has the facility to record and replay programs onto and from standard audio cassettes, or to transfer a program to an erasable programmable read only memory (EPROM) via the EPROM writer. Once programmed, the EPROM chip

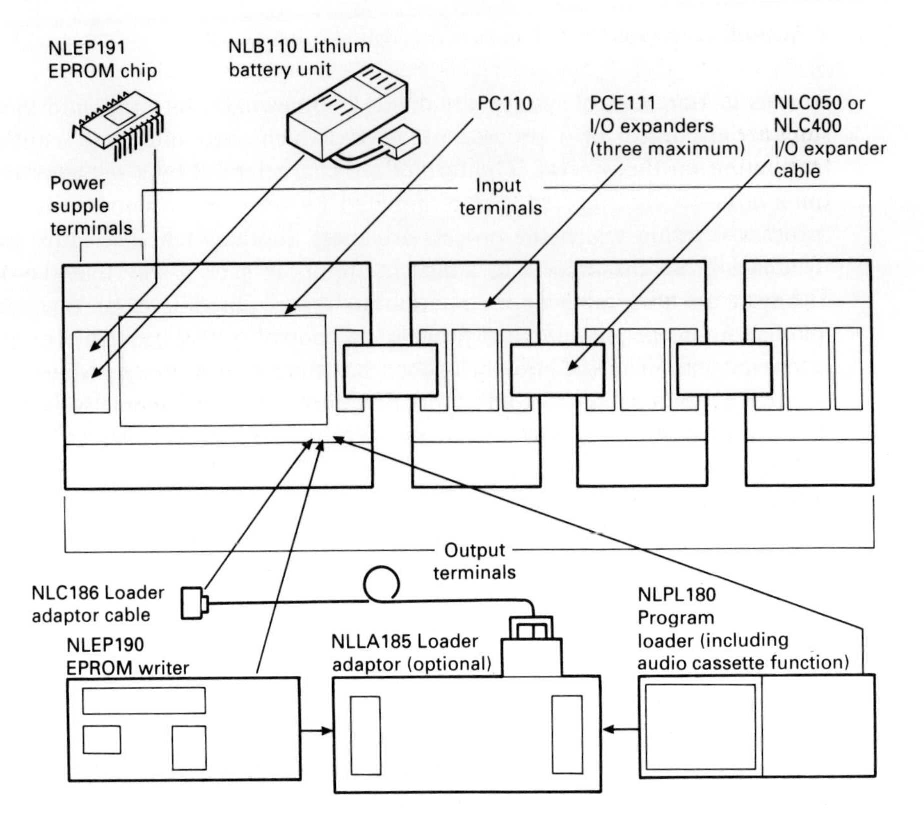

Figure 2.7 Programmable logic controller (Westinghouse).

retains its information indefinitely and when plugged into the main unit its contents are transferred to the process memory on start up. The lithium battery provides extended back-up power for the memory (up to 3 years).

Commonly, the input and output terminals interface with the CPU via optoelectronic isolators designed to protect the latter from electrical interference which could cause faulty operation or damage to the system. At the input a simple contact closure (indicating *on* or logic 1) is all that is needed to switch on an LED which is powered by an internal voltage source. That illuminates a photodiode, which passes on the signal to the CPU. Triggering from voltage-free contacts in this way is to be preferred. However, other sources of logic 1 inputs are sometimes unavoidable and it is possible to derive them from a.c. mains contacts or from analogue signals when they attain a pre-selected level.

System integrity

Optoelectronic isolation is one measure taken to maximise the integrity of the PLC. It must also be able to withstand a wide range of ambient temperatures and humidities, as well as high dust levels and some mechanical abuse.

Manufacturers normally quote electrical noise immunity to a specified standard, such as IEC 255-4. Nevertheless, they expect their PLCs to be sited away from high voltage, high power control panels and that noise suppression measures are applied to local sources such as large inductive loads. Input and output terminals are usually rated to withstand 1500 V a.c. relative to their mountings for 1 minute. Vibration limits and flame retardancy are also quoted by some manufacturers.

Internally, the processor's logic program should also carry out diagnostic tests. The system depicted in Fig. 2.7 checks at start up that all system hardware is functioning properly. A checksum calculation is performed on the memory and compared with the result obtained when the processor was last powered down, to ensure that memory back up has been effective. When all of these tests have been successfully completed the processor will start to solve logic and service the inputs and outputs.

When the processor is running it continues to test hardware and user software for errors. A watchdog timer is also incorporated to guard against processor failure. While the processor is running a RUN contact is closed. If any error is detected this circuit opens and all outputs are disabled.

Error codes are displayed by the program loader in the event of an incorrect keystroke or an equipment malfunction.

Programming facilities

The computer memory available to the user is divided into two sectors – the program memory which holds a set of program instructions or steps, and the data sector which holds information on the current states of the inputs and outputs, together with a software representation of the required sequential control system.

The organisation of the memory in the PLC of Fig. 2.7 is given in Table 2.3, which shows that it accommodates a maximum of 1023 program steps and has 64 eight-bit registers in the data memory. The address of any point is defined by its location within a register – for example, the addresses of the points in the first register are 0.0 to 0.7. The logic status (off or on) of each input and output terminal is registered at its address as 0 or 1, respectively.

The free registers hold the software representations of internal contacts, coils, timers, counters, etc, created by the program instructions. In particular, the first seven bits of a register programmed to act as a timer are used to count down from a preset number in 0.1 s decrements, using the computer's internal 0.1 s timer impulses. In this way a time delay up to $127 \times 0.1\ s = 12.7\ s$ can be created before the countdown reaches zero. At that point the register's eighth bit changes state to provide the timer's output signal. This can be used to start the countdown in another register if a longer time delay is needed. Counting of external time signals or other events is achieved in the same way.

Programming

The earliest form of PLC programming – still widely used – is based on the ladder

Table 2.3 Programmable logic controller memory organisation (Westinghouse).

	Type	Allocation	Address	Remarks
Program memory	Program instruction	1023 steps	Step numbers 0–1023	Capacitor back up Additional back up by optional lithium battery
Data memory	Inputs (PC110)	24 points	0.0–2.7 (IN1–24)	
	Outputs (PC110)	16 points	3.0–4.7 (CR25–40)	Unused I/O addresses can be used as additional internal data addresses
	Inputs (PCE111 I/O expander 1)	16 points	5.0–6.7 (IN41–56)	
	Outputs (PCE111 I/O expander 1)	8 points	7.0–7.7 (CR57–64)	
	Inputs (PCE111 I/O expander 2)	16 points	8.0–9.7 (IN65–80)	
	Outputs (PCE111 I/O expander 2)	8 points	10.0–10.7 (CR81–88)	
	Inputs (PCE111 I/O expander 3)	16 points	11.0–12.7 (IN89–104)	
	Outputs (PCE111 I/O expander 3)	8 points	13.0–13.7 (CR105–112)	
	Internal data including internal contacts and coils	34 registers (272 points)	14.0–47.7	Volatile data is lost when power is turned off
	Timers Counters Shift registers Step controllers	16 registers (128 points)	48.0–63.7	Non-volatile data is retained when power is turned off

network. This representation of the required system was developed to aid the design of traditional relay systems and is known as relay ladder logic, or RLL. The designer's ability to employ a familiar technique greatly facilitated the adoption of the PLC.

Figure 2.8 provides an example of a ladder diagram, showing its structure and the type of symbols employed. The convention is that the topmost rung of the

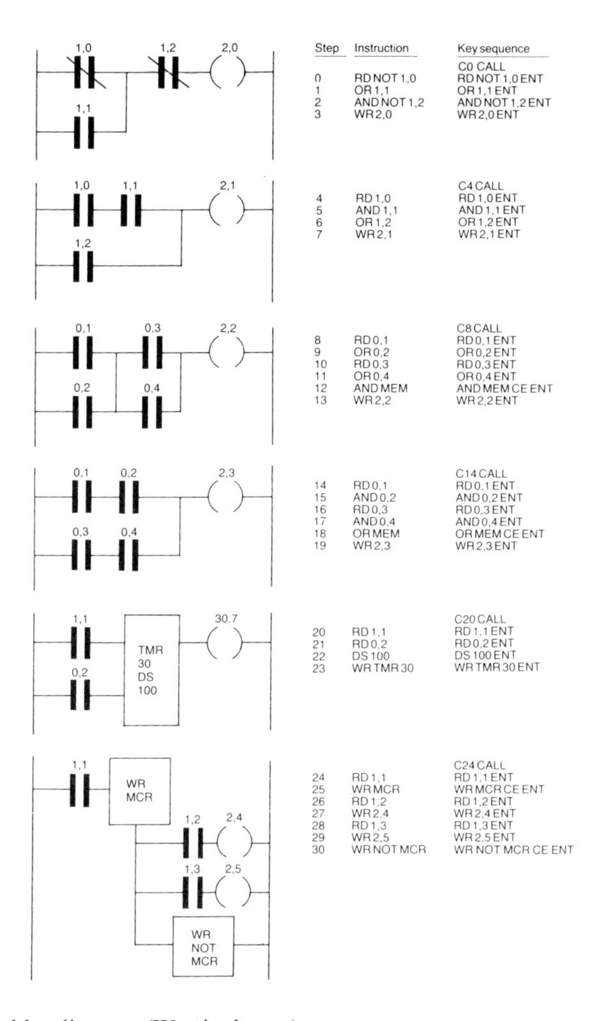

Step	Instruction	Key sequence
		C0 CALL
0	RD NOT 1,0	RD NOT 1,0 ENT
1	OR 1,1	OR 1,1 ENT
2	AND NOT 1,2	AND NOT 1,2 ENT
3	WR 2,0	WR 2,0 ENT
		C4 CALL
4	RD 1,0	RD 1,0 ENT
5	AND 1,1	AND 1,1 ENT
6	OR 1,2	OR 1,2 ENT
7	WR 2,1	WR 2,1 ENT
		C8 CALL
8	RD 0,1	RD 0,1 ENT
9	OR 0,2	OR 0,2 ENT
10	RD 0,3	RD 0,3 ENT
11	OR 0,4	OR 0,4 ENT
12	AND MEM	AND MEM CE ENT
13	WR 2,2	WR 2,2 ENT
		C14 CALL
14	RD 0,1	RD 0,1 ENT
15	AND 0,2	AND 0,2 ENT
16	RD 0,3	RD 0,3 ENT
17	AND 0,4	AND 0,4 ENT
18	OR MEM	OR MEM CE ENT
19	WR 2,3	WR 2,3 ENT
		C20 CALL
20	RD 1,1	RD 1,1 ENT
21	RD 0,2	RD 0,2 ENT
22	DS 100	DS 100 ENT
23	WR TMR 30	WR TMR 30 ENT
		C24 CALL
24	RD 1,1	RD 1,1 ENT
25	WR MCR	WR MCR CE ENT
26	RD 1,2	RD 1,2 ENT
27	WR 2,4	WR 2,4 ENT
28	RD 1,3	RD 1,3 ENT
29	WR 2,5	WR 2,5 ENT
30	WR NOT MCR	WR NOT MCR CE ENT

Figure 2.8 Ladder diagram (Westinghouse).

ladder represents the first operation in a control sequence and subsequent operations are represented by lower rungs in a logical sequence. At each rung the contacts that are involved in the initiation of the operation are placed on the left side of the rung and the corresponding output device is on the right, again by convention. When a conducting path through the contacts has been established the output device will be activated. When used with a PLC the program instructions must also follow this order.

On the first rung in the diagram the output device which is linked with the memory address 2.0 will receive a *run* signal from that address until either of the two normally closed (NC) contacts, 1.0 and 1.2, opens – i.e., when either address receives an *on* signal from its corresponding input terminal. If NC contact 1.0 opens, the closure of normally open (NO) contact 1.2 will restart the output device.

On the second rung contact 1.0 is shown as NO (evidently it is a changeover switch). The contacts 1.0 and 1.1 are in a logic AND configuration since both have to close before output device 2.1 can operate. The alternative path via contact 1.2 provides an example of the logic OR configuration. The third rung has two ORs in an AND configuration while the fourth rung has two ANDs in an OR configuration. These ANDs and ORs can be seen in the corresponding list of instructions in column 2 of Fig. 2.8. The four program steps needed to enter the first rung of the latter (first sequence) also show that in this instruction set the NC condition is described by NOT. The initial RD (read) at step 0 and the final WR (write) at step 3 define the start and finish of the instructions for a rung. The final column shows the key sequence to be entered through the keypad/display unit, which allows the programmer to view the instructions before ENT (enter).

[*Note* Programming detail still varies from vendor to vendor, but the industry is closely involved in development of the IEC 1131 standard for PLC programming.]

The fifth rung shows a timer at register address 30, which has been programmed to produce an output after it has received 100 time pulses (DS = data set). In this case the eighth bit of the register (address 30.7) is the output device, as explained earlier. The software provides two inputs, or gates, to the timer. The uppermost is the enable/reset gate which activates the timer when contact 1.1 closes and resets it to its start count (100 in this case) when 1.1 opens. The second is the time/hold gate which allows the timer to count down when contact 0.2 is closed, or to hold the current count when it is off. Countdown is resumed if it returns to on, provided that the upper gate is still in the enable mode. The output at 30.7 is off until the end of the timing period, when it switches on and remains at that level until the upper gate resets the timer.

In effect 30.7 acts as an internal contact which can be used as an input to another rung of the ladder. This can be on a higher or lower rung than the timer, which exemplifies the point that the order in which the inputs are introduced in the ladder does not necessarily coincide with the order in which the outputs are arranged. In practice, this does not matter because in operation the CPU repetitively and rapidly scans the inputs to determine their status, then scans the program before adjusting the outputs as necessary. The standard measure of scan speed is ms/kbyte. The scan time depends on the length of the program, of course, but with scan speeds of under 1 ms/kbyte now available, the PLC is capable of controlling quite rapidly changing processes.

Another commonly used programming method employs flow chart symbolism and instruction sets. This method is often preferred for set sequences of operations. Some sequential control systems incorporate both ladder and flow chart methods. Programming of larger systems is also facilitated by PC-based software, providing graphics which allow the user to construct and modify a ladder or flow chart representation of the system interactively. The program can be developed, tested and printed out with the PC connected to the PLC via the former's RS232 serial port, or programming can be done off-line and the program loaded into the PLC when required, either by the same link or via a floppy disk.

Applications of the PLC

Many of the PLC's uses are for sequential control of single items of equipment, including domestic washing machines. Individual PLCs can also control the sequence of operations in a materials handling plant. An example is given in Chapter 5. Large-scale SCADA systems have distributed PLCs (see Fig. 2.5) which communicate with clients and each other over one or other of the networks already described. It is also possible to use a multidrop serial communications link such as RS485 to separate the physical input and output connections from the PLC, leaving the main wiring at the field sites and requiring only a single power/signal cable connecting these sites to the PLC.

Many PLCs now provide additional functions, normally associated with continuous control rather than sequential control. This has come about because the CPU is capable of processing analogue inputs and outputs digitally, via the A/D and D/A converters mentioned in the first section of this chapter, at sub-millisecond conversion rates. In particular, these hybrid units are often applied where closed loop control of temperature and flow is part of the system requirement.

In fact, the gap between the PC and the PLC has continually narrowed and since the early 1990s PCs which use a special operating system to emulate PLCs have been available commercially.

Fuzzy logic

This topic is introduced here in the context of the PLC because some PLC applications incorporate this form of logic. However, fuzzy rules have widespread industrial applications, not least in agriculture, and the topic arises again later in this chapter, within the context of artificial intelligence.

Essentially, the need for fuzzy logic arises when there is no clear-cut relationship between the value of the inputs to a system and their influence on its output. It deals with uncertainty and imprecision, which calls for the application of fuzzy set theory, introduced by L.A. Zadeh in 1965. In the world of fuzzy sets a data input is assigned membership to one or more of a set of overlapping membership classes, or membership functions (MSFs) which are defined by linguistic variables, rather than numerical ranges.

Figure 2.9(a) shows three possible membership classes of a quantity, linguistically defined as *low*, *normal* and *high*. The vertical axis provides the *membership value*, *truth value* or *degree of association* of the data input with respect to the membership classes. The value 1.00 indicates that it belongs absolutely to that class and the value 0.00 indicates that it is absolutely not a member of the class. According to the figure the variable will be absolutely in class LOW up to a value of 20 and absolutely not in that class beyond the value 26, with intermediate degrees of association between 20 and 26. From 20 to 26, it also has increasing

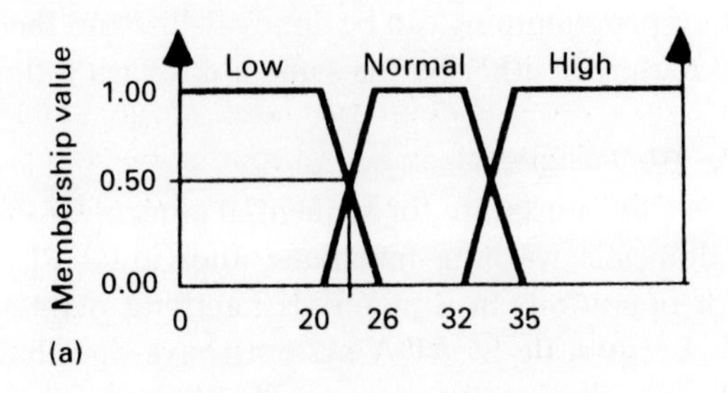

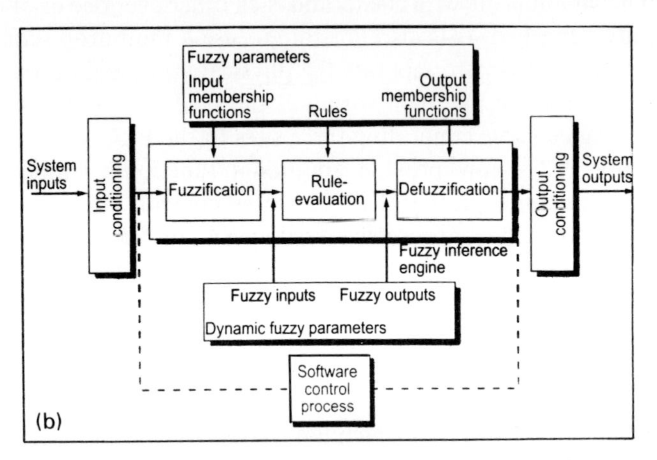

Figure 2.9 (a) Fuzzy sets; (b) fuzzy logic control system (EPD).

degrees of association with class NORMAL, and so on. In this example the membership classes are shown as trapezoidal in shape, which is a common representation. A narrower flat top and steeper sides indicate greater certainty about the input data – tending to a narrow triangle with its apex at the midpoint.

Figure 2.9(b) shows the three main elements of a fuzzy logic control system. *Fuzzification* is the process of assigning data inputs to the defined membership classes of their variable, with the appropriate membership values. At the next stage groups of fuzzified inputs are submitted to a set of fuzzy IF-THEN rules, based on human knowledge and judgement. When these rules are *fired* by the built-in program (rule evaluation) they provide a set of fuzzy outputs which are combined and then converted to the defuzzified (crisp) output that is required for control action. The rules can be weighted according to their estimated importance.

To illustrate this sequence, control of room ventilation might be based on inputs generated by internal and external air temperature and RH sensors, an anemometer providing wind speed and direction, and a solarimeter. These might provide inputs to fuzzy sets with MSFs low, medium and high or, in the case of wind direction northerly, southerly, etc. A variety of rules might be devised to evaluate the need for ventilation. For example, if the internal air temperature is high AND the wind speed is low AND the wind direction is southerly THEN ventilate. Firing of all the rules such as this would lead to a set of fuzzy outputs related to the quantity ventilation. These would then be combined (after due weighting) into a single defuzzified output by one of the methods that have been developed for this purpose. The *centre of gravity* method is commonly used. This can be applied in the following way. If the set of fuzzy outputs is presented in the form of a set of n membership values (y, between 0 and 1) at position x on the scale of ventilation, then the system output can be calculated from the ratio of two sums:

$$\frac{\sum_{i=1}^{n} x_i \times y_i}{\sum_{i=1}^{n} y_i}$$

Specific examples of the application of fuzzy logic appear later in this book. Here it can be noted that many standard commercial temperature controllers employ fuzzy logic to improve their settling time at start up and after set-point changes and other disturbances. Embedded fuzzy logic chips are also widely used in the operation of camcorders, automatic transmissions and, once again, washing machines. Some chips also provide the facility to modify the midpoint and width of any membership function dynamically. This facility can be use to adjust the drift of sensors, *inter alia*.

Closed loop control

It has been mentioned that the analogue inputs and outputs incorporated in some PLCs are commonly used for automatic control of temperature and flow. This usually implies closed loop control, as shown diagrammatically in Fig. 2.10. The measuring instrument, containing a sensor or sensors, monitors the state of the process, thereby providing a measure of the controller's success in achieving the control setting. Any disparity is fed back as an error signal to the controller, closing the loop. Corrective action ensues. In the absence of such a feedback loop the process would need to be so predictable that a given control setting would effectively guarantee the required condition of the process. Many processes are subject to too many external disturbances for this (open loop) control to be satisfactory.

So much is straightforward but the dynamics of closed loop controls are not and this has given rise to the extensive literature on control theory. Control theory applies mathematical analysis to the dynamic response of the elements of a control system and to the stability of the system's response at different frequencies. It also enables the designer to take account of the effect of external disturbances (noise) on the system.

As with the sensor dynamics of Chapter 1, the reader is referred to standard

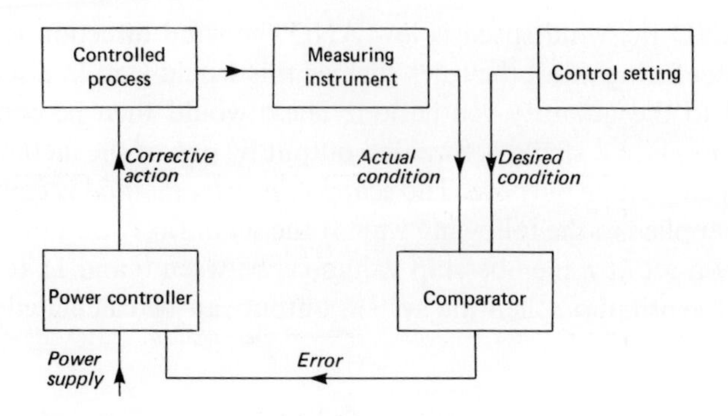

Figure 2.10 Elements of a closed-loop control system.

text books for details of these analytical techniques. Tutorial software is also available from suppliers of mathematical suites for computer-based analysis and design. Here somc of the more basic concepts are outlined.

Algorithms

The control algorithm built into the controller determines the corrective action that results from the error signal in Fig. 2.10. Its nature affects the closeness of control that can be achieved in slowly or rapidly varying conditions.

The most elementary form of control is the on–off mode, shown in Fig. 2.11(a), which could relate to a simple thermostat, controlling the heat input to a process. Here the algorithm amounts to 'switch off when the measured temperature reaches level A and switch on when it falls to level B'. The differential is necessary to avoid the state of indecision that would occur if there was a single threshold but it immediately places a limit on the closeness of control that can be achieved. In addition, it cannot avoid a measure of overshoot in each direction, because the heating system will have a finite response time and there will be a delay while the effect of the change in the heat input is registered by the thermostat, wherever it is situated. The latter would be reduced by placing the sensor closer to the heat source but, of course, that would increase the likelihood that the measured temperature was not representative of that in the heated system.

Liability to overshoot the set point increases with the rate of input in the *on* phase, but too low a rate increases liability to overshoot in the other direction. The rate that gives the best compromise depends on the individual application. Similarly, the setting of the differential band is a compromise between the rapid and possibly disadvantageous switching (*hunting*) which can result from a narrow band setting and the less precise control which accompanies a wider one.

The PID algorithm was introduced to overcome the limitations of on–off control, and has become a common standard for environmental control, long since adopted in heated glasshouses. This applies correction in proportion to the size of the error, to the integral of the error, and sometimes to the rate of change (or derivative) of the error. This is known as two-term (PI) or three-term (PID)

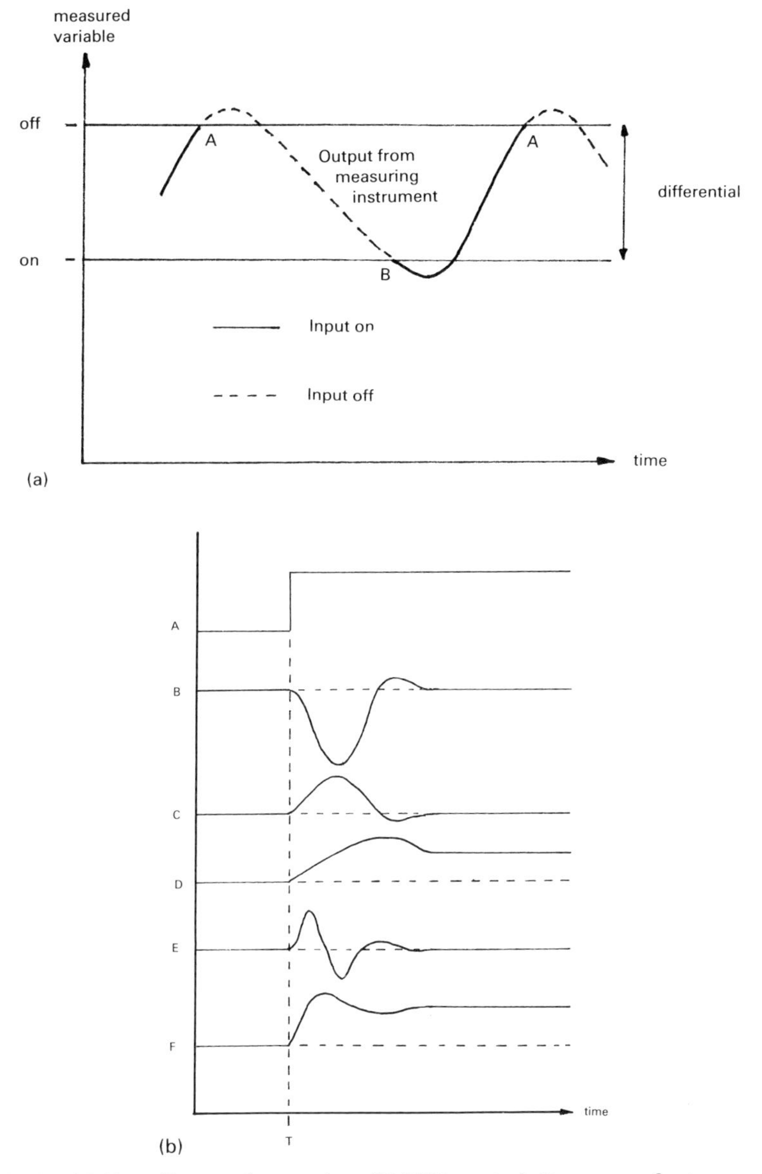

Figure 2.11 (a) On–off control operation. (b) PID control. Response of a temperature controller to a sharp change in the required heating load (A) at time (T). B – resulting temperature change. C, D, E – contributions of the P, I and D terms to the changed position of the heating valve. F – resulting change in the valve's position.

control, depending on the presence or absence of the derivative term in the control action. The contributions of the P, I and D terms to this form of action can be seen in Fig. 2.11(b), which shows the effect of a sudden increase in the heating load on a room temperature controller, perhaps caused by a sharp drop in the outside temperature. The heat input is controlled by a valve such as that found in domestic heating systems. To increase the heat input and so restore the air temperature to its set point the valve must open to a new setting. The contributions to the opening of the valve by the P, I and D terms individually and collectively are shown in the figure. Proportional action can never restore the temperature to its set point because it opens the valve by an amount proportional to the error and then seeks to return to its original as the error diminishes. The valve thereby settles at a position in which there is a residual error but further reduction would reverse the temperature rise and so increase the error again. The resulting residual error is known as the proportional *offset* or *droop*. Integral action overrides this offset by continuing to open the valve while there is an accumulated error. Derivative action helps to shorten the time taken for the valve to reach its new equilibrium position. Several forms of integral action are commonly employed, including a *reset* action which applies a constant correction as long as an error exists. All have the same aim, i.e., to bring the temperature back to its set point after a disturbance as quickly as possible and with minimal overshoot. The appropriate magnitudes of the P, I and D terms depend on the particular applications, therefore two- and three-term controllers are provided with adjustments for *tuning* the complete loop.

Self-tuning is now a feature of many PID controllers. This enables them to adapt automatically to disturbances in the process or changes of the set point, with consequent reduction in overshoot and settling time, relative to occasional manual tuning. Some respond to the effects of the above disturbances on the process – i.e. *after* the event. Others incorporate fuzzy rules, to make changes as soon as a disturbance is sensed. The latter can provide much tighter control than the former.

The genetic algorithm (GA) is an altogether different concept. It brings the mechanisms of genetic selection to the task of finding an optimal value for the set point in complex processes, such as optimisation of plant growth environments. This concept is outlined in conjunction with that of the neural network, which follows here.

Artificial intelligence (AI)

Increasingly, industrial control systems incorporate a measure of AI, made possible by the power and speed of computers and their peripherals. They provide the means to hold complex processes at optimal performance, whether at overall plant level or at subordinate levels. Major industries employ these techniques to maximise output in terms of quantity and quality, and to monitor machine health for early warning of impending component failure. These are beginning to find application in the agricultural sector – particularly for quality analysis and control – as will be shown in later chapters.

Complex processes may be regulated by hierarchical combination of expert systems, artificial neural networks, fuzzy controls and genetic algorithms. Fuzzy logic control has been covered earlier, the other three components will be dealt with in turn.

Expert systems

The expert system dates from the early 1980s and has been widely employed as a decision aid, in agriculture as elsewhere. It is essentially a knowledge-based system (KBS) which holds facts, rules and judgements related to a particular domain, coupled with an *inference engine* which interprets the knowledge to provide diagnoses and prescriptions in the domain area. In the control context it is found at the top level of an hierarchy, providing overall steering of the system. Like fuzzy logic control, it is dependent on human expertise and reasoning to provide the foundation for its operation.

However, there are many domains in which specialist expertise is sparse, usually due to their complexity. In these cases the artificial neural network (ANN) offers a solution.

Artificial neural networks

The neural network is a piece of software that mimics the architecture of the human brain and it operates in somewhat the same way, although on a much smaller scale at present. Its prime function is to act as a trained pattern-recognition device. Like the brain it is in the form of a network of interconnected neurons or nodes which gather input information and contribute collectively to an output decision based on prior experience. In fact it is an example of parallel processing of information, as opposed to the serial (step-by-step) processing of the traditional computer.

Most ANNs contain only a few neurons, as indicated above. An example of a three-layer network is given in Fig. 2.12. The input and output layers are separated by a *hidden* layer, in which all the nodes are connected to all of those in the other layers. The number of nodes in the input and output layers is defined by the application, while those in the hidden layer are usually determined empirically.

The four inputs in Fig. 2.12 supply binary data on representative features of an object or data set. The four representative features of an agricultural object might be its degrees of greenness, yellowness and redness, and its weight. The objective is to train the network to allocate sets of input data to one of the four classes corresponding to the four outputs. In the above example these might represent *large ripe*, *small ripe*, *large unripe* and *small unripe*.

The training process employs sets of input data which are known to possess the characteristics of each output class. In passage from the input nodes to the hidden layer each signal is given a weighting. Then the sum of the weighted inputs at each hidden node is passed to the output nodes after a further weighting by an *activity*

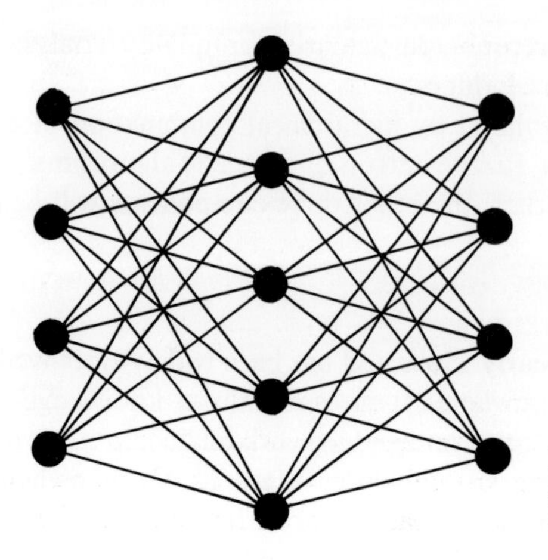

Figure 2.12 Three-layer neural network with four inputs and outputs, and a hidden layer with five nodes.

function. The latter is usually in sigmoid form, which progressively attenuates larger signals. Initially the computer allocates random weightings to the input signals, which almost invariably leads to incorrect outputs. The resulting errors are then *back propagated* from the outputs, to adjust the input weightings iteratively in a systematic way, until the required classification is attained with an acceptable level of reliability. In general, many training sets are employed for this process and thousands of iterations may be required. Then the network is presented with another set of representative inputs, to determine the degree of confidence that can be placed on its ability to classify unfamiliar inputs. When that test has been satisfied the network is ready to classify unknown data sets.

In many applications there is only one output but there can be ten or more inputs. There can also be more than one hidden layer but most applications relevant to the agricultural sector have been based on the three-layer model. Although the size of that layer bears some relationship to the complexity of the classification problem there are no clear rules by which the required number of hidden nodes can be estimated. Common practice therefore is *either* to start with a few and determine whether improved performance is obtained by adding to their number, *or* to start with an unusually large number and to reduce it until performance falls below the acceptable level. In any event, commercial software is available for the design and application of individual networks.

In the control context, as opposed to the classification of inputs, the ANN can be used in another way. If the state of a process is sampled at regular intervals to create an historic time series of that state, and if the inputs to the process are simultaneously sampled, an ANN can be trained to provide a computer model of the process, relating the output to the inputs. The model is derived by applying corresponding sets of these input and output data to the input nodes, simultaneously. The resulting output at the single output node is then compared with the value required by the time series. Any difference is treated as an error to be eliminated by the iterative back-propagation technique already referred to. At the end point the weightings in the ANN have been adjusted to provide the required input/output model. The ANN is then able to predict the output from any set of inputs. This provides a means to determine the set of inputs which will provide optimal control of the process.

Genetic algorithms

Search strategies are often required to find the set points for optimal control, but there is always the possibility that they will converge on a suboptimal maximum. However, genetic algorithms inherently search *solution space* widely and so are less likely than others to do this. In consequence, they are regarded as a particularly efficient search tool. They are attributed to the work of J.H. Holland, dating from the mid 1970s, and have been employed in whole-farm planning and dynamic simulation modelling of agricultural systems. Just as ANNs are related to the mechanics of the brain, GAs are based on the mechanics of natural selection.

The search process starts with the generation of a population of *chromosomes* in the form of bit strings (chains of, say eight 0s and 1s), each representing a possible candidate for the optimal control setting. These binary strings are selected randomly. The chromosomes are then subjected to evaluation by a fitness function which identifies those with the most useful properties. This test can be carried out by making each one the input to an ANN process model such as that just described. Those chromosomes that provide the highest outputs from the ANN are the fittest candidates for the optimal set point.

The second stage is the selection of pairs of the fittest chromosomes for reproduction. *Offspring* are generated by manipulation of the *genes* (the individual bits) in the chromosomes. The two most common of these operations are *crossover* and *mutation*. These are illustrated in Table 2.4. Cross-over is effected by exchanging a randomly selected segment between two chromosomes, while mutation results from the change of a randomly selected bit in one of the pair, from 0 to 1 or vice versa. These manipulations are carried out probabilistically and in a way that creates far more crossovers than mutations.

Table 2.4 Genetic algorithms: (a) exchange of *chromosomes*; (b) mutation.

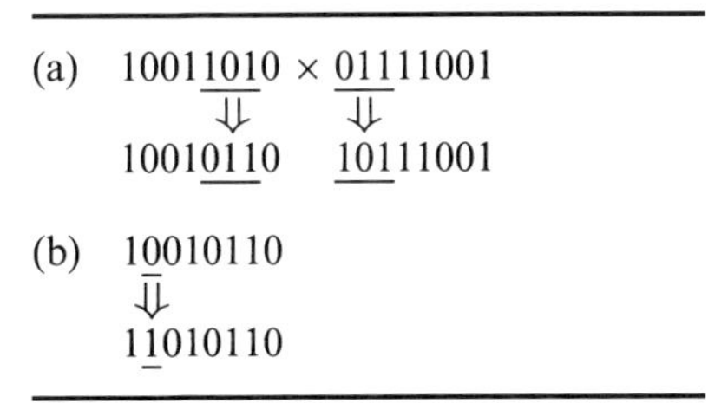

(a) 10011010 × 01111001
⇓ ⇓
10010110 10111001

(b) 10010110
⇓
11010110

The new population is then tested for fitness and the fittest again survive, to be paired and to create more offspring. The cycle continues until the average fitness of the selected population converges to the point at which it can be treated as a near-optimal set point. The intervening wide search of solution space results from the variations produced by the genetic operators (cross-over and mutation).

Machine vision

The last AI topic that will be introduced here concerns the methods of feature extraction from images of various kinds, in order to monitor and control processes including quality determination and robot operations. These are usually classified under machine vision or computer vision. There are four common elements in most vision systems:

- An imaging device which generates a two-dimensional representation of a scene or object. This is commonly the CCD camera introduced in Chapter 1 and its output feeds into the computer which performs the subsequent operations.
- Segmentation and thresholding of the image, to select areas of interest in it and to present them in binary (pixel) form. Depending on the application this can be done on a grey scale (monochrome) or on the basis of colour.
- Feature extraction, to determine shapes, areas or colour variations, etc. in the selected areas. This phase employs a variety of search algorithms for specific purposes, such as boundary mapping.
- Classification, which frequently involves neural networks trained to recognise the feature(s) of interest.

Examples of these techniques will be found in later chapters.

Output devices

Agricultural control systems provide applications for a wide variety of electrically powered actuators and regulators but the most widely used is probably the electric motor in one of its many forms. Among these the a.c. induction motor, for either single- or three-phase operation, has long been a familiar power source in many types of installation on farms of all descriptions. In the context of this book two types of intelligent control for these ubiquitous devices deserve mention at this point.

Soft start

The soft start principle is illustrated in Fig. 2.13. The current in one of the three phases of the motor supply is sensed to provide a feedback signal. As shown in the diagram, the sensor is of the inductive type, akin to portable, clamp-on ammeters which do not interrupt the wiring. A microcontroller then applies a regulating algorithm to three sets of semiconductor switches known as thyristors or SCRs (silicon controlled rectifiers), which control the proportion of each half-cycle of the a.c. supply in each phase equally. That regulates the average or rms (root mean square) voltage to the motor.

The initial purpose of this system was to avoid the heavy inrush current that

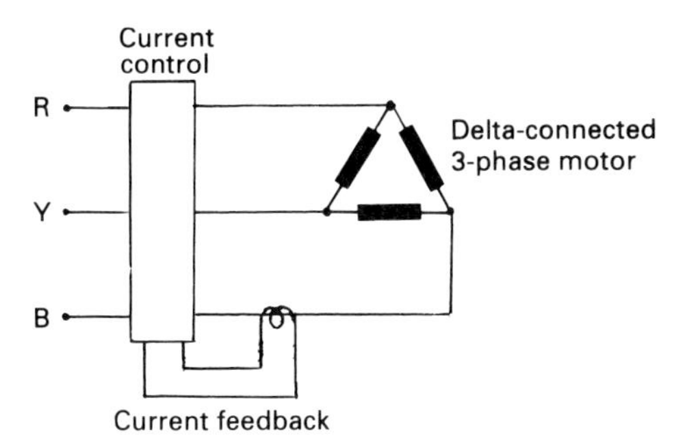

Figure 2.13 The soft start principle (Farm Electric Centre).

occurs when an induction motor is first switched on, because the resulting peak power demand can add to the capital cost of the installation in addition to its running costs. The controller achieves this by applying a ramp function to the thyristors in order to bring the current up to its operational level at a controlled rate. The ramp can usually be adjusted in the range 1 second to 1 minute, as required.

This principle is not appropriate for all motor drives, since the resultant motor torque (proportional to the square of the applied voltage) is low at start up (low speed). However, where the torque requirement is low at low speeds – as it is in many fan and pump installations – the soft start is advantageous. The controller can also be used to sense light loads (reduced current) and to reduce the motor's voltage input accordingly, thereby reducing *overfluxing* (unnecessarily high magnetising current) without reduction of performance. This can make significant savings in energy consumption through reduced motor losses and improved power factor (PF).

An option also found in these control units is a soft-stop function, which reduces mechanical stresses in the drive systems for pumps, conveyors and other devices at turn off, by ramping down the current. A complete soft-start unit will also have self-diagnostics (thyristor open-current or short circuit, for example) in addition to standard contact breakers for dealing with electrical and thermal overloads.

Variable speed drives

Inverters have been available for a.c. motor control for many years and for a wide range of applications, including fan, pump, compressor, conveyor and position control. Each application presents obstacles to smooth and energy-efficient operation, for example, compressors require high starting torque; heavy duty conveyors may be continuously loaded or shock loaded; fans may run inefficiently at times, as indicated above. In the past, ways of addressing these problems required bulky and expensive equipment. However, developments in microelectronics have led to the design of compact, multifunction units which convert mains frequency power to a variable frequency power supply, typically in the range from 60 or 50 Hz down to about 0.5 Hz, with considerable precision and with the ability to match the torque output of motor drives to their instantaneous requirements.

For switching of power below 500 kW the most common switching element is no longer the thyristor but the power transistor, best known as the isolated gate bipolar transistor (IGBT). This is usually switched by a software pulse width modulation (PWM) generator, which synthesises three-phase sine waves from pulses of cyclically varying width and with repetition frequencies of about 15–20 kHz. The variable frequency power produced by these units can be controlled very precisely and the above audiofrequency of the PWM generator reduces the acoustic noise that can be generated in the motor, due mainly to magnetostriction effects in its laminations. However, the same high PWM frequency can cause problems of radiofrequency interference (RFI) which require suppression to satisfy legislative requirements.

The next step is to measure the motor torque in some way at any speed. This has led to the introduction of flux vector control. Vector drives control the lines of magnetic flux between the motor's stator and rotor with the aid of a computer model of its electrical characteristics over its working range. These characteristics, programmed into the computer's memory as a look-up table, combined with data on the voltage, current and frequency being generated by the inverter, are the basis for sensorless vector control. The addition of *sensorless* distinguishes the system from the earlier, closed loop form which obtained details of the rotor's speed and angular opposition from an encoding disc on the shaft. That system simplified the calculation of the inverter's settings for optimal torque control, and it allowed delivery of full torque at zero speed. Nonetheless, it has been displaced by the open-loop sensorless system, which manages without an encoder and can typically deliver 250% of full load torque at 0.5 Hz. To achieve this it has to be programmed with details of the motor being driven by the inverter (type, rated power, voltage and frequency). Then it performs a tuning (or learning) run with the motor at the outset, in order to memorise its characteristics.

Some autotuning systems include the mechanical inertia of the motor drive in the stored characteristics, to aid smooth, trip-free running. Fuzzy logic is also employed as a means to look for trends in the operation and to initiate adjustments without causing tripping. Energy management includes the use of excess energy for regenerative braking and in some cases the feedback of electrical energy into the mains network.

Diagnostics include protection against short circuits and open circuits, earth faults, overcurrent and overvoltage, motor stall, motor and inverter over-temperature. Some units can model motor temperature from their data on current, voltage and frequency and can accommodate the different characteristics of fan-ventilated motors, to protect them in accordance with their specified temperature rise. Others have the facility to monitor motor temperatures directly, and this information can be used to improve vector control itself. Overall, the design aim is to avoid damage to the system through early warning measures coupled with controlled shut down when this is essential.

The computing requirement for the sensorless system is very demanding and smooth regulation requires rapid response to changing demands. Therefore the computing is generally handled by a digital signal processor (DSP) rather than a conventional microprocessor. The DSP is designed for the heavy duty arithmetical calculations associated with fast-reacting, real-time control of processes.

All of these drive units are capable of remote control and diagnosis via the standard communications links described in this chapter.

The man–machine interface (MMI)

Despite the advances made by automatic control, including robotics, people still need to interact in many ways with installed instrumentation and control systems. The MMI is therefore an element of these systems that needs to be considered. Ergonomics and safety factors can be important in this context, especially when swift action is required.

Displays need to be both visible and legible from the operator's working position. They should avoid information overload that can detract from the operator's monitoring of key parameters, particularly those that may call for a quick response. A display of the emergency action required adds to the value of an alarm signal.

The siting and mode of operation of controls should be based on the ergonomics principles which have been well documented over many years. The siting and accessibility of emergency stops is, of course, crucial, as is conformity to electrical safety standards in the whole system.

The value of initial training in the use of new equipment should be recognised. Operators left to learn by experience after only a brief introduction to the equipment can make avoidable and sometimes costly mistakes. A well-prepared instruction booklet, with clear illustrations and simple diagnostics, can reduce misunderstanding and emergency calls to the supplier.

2.4 System integrity

Instrumentation and control systems – particularly if they were electronics based – were once regarded as too costly and unreliable for general use in agriculture, as well as marginal in their economic value to most farming activities. The pressures to achieve greater efficiency, improved quality control and environmental friendliness have made computer-based systems increasingly relevant to management and control of many aspects of crop and livestock production. Because these systems are built from mass-produced, quality controlled components their cost is increasingly affordable. System reliability has undoubtedly increased too, partly because so many of the components are the quality-controlled microcircuit chips, and partly because most manufacturers of instrumentation and control equipment for the agricultural sector have learned what it takes for this type of equipment to survive in that environment. System software has become more stable, too.

Nonetheless, vulnerable areas remain. In addition, there is now a legal requirement for the equipment itself to be environmentally friendly by avoidance of the emissions to which references have been made in this chapter. These considerations place some responsibilities on the designer and some on the user. They will be outlined in turn.

Design requirements

Within the European Economic Area (EU and EFTA) CE marking of products was introduced to allow free movement throughout the Area. Products must comply with appropriate essential requirements, which can be conformance to a relevant European standard. Manufacturers of equipment must bc able to provide a Declaration of Conformance to back up the CE marking of their product. Directives (i.e. Community law) relevant to instrumentation and control systems have been issued, including:

- The Machinery Directive, operational from January 1995. This includes mechanical, electrical and worker safety requirements, excluding the electrical aspects covered by the Low Voltage Directive, below.
- The Electromagnetic Compatibility (EMC) Directive, operational throughout the EEA from January 1996. This covers the range of phenomena and effects listed in Table 2.5. [*Note* Spark-ignition engines of tractors and other vehicles are covered by other directives.]
- The Low Voltage Directive, operational from January 1997. This covers all components using voltage ranges from 50–1000 V a.c. and 75–1500 V d.c. It is concerned with electrical hazards to 'persons, domestic animals and property', together with the mechanical and electrical aspects of their protection.

For these reasons manufacturers quote IEC standards or European norms (ENs) for EMC compliance, and IP numbers for the physical protection offered by enclosures. Electromagnetic compatibility compliance concerns both the emission of radio frequency interference (RFI) by equipment and its immunity from RFI, either conducted to it by mains and other cables or radiated to it. For example, EN 50082.2 covers generic immunity in industrial environments and specifies protection in terms of the attenuation of RFI in decibels (dB), the common logarithmic scale for amplification and attenuation of electrical signals. The specification for mains-conducted RFI covers the frequency range of 0.1 to 100 MHz and that for radiated RFI covers 10 to 1000 MHz.

It should be noted that the high-speed switching of digital signals characteristic of contemporary computers generates plentiful radiation of these high frequencies and that it requires enclosures of high screening integrity to achieve

Table 2.5 European EMC directive: Phenomena and effects which may be regarded as electromagnetic disturbance.

(1) Conducted low-frequency phenomena:
- slow variations of supply voltages;
- harmonics, interharmonics;
- signalling voltages;
- voltage fluctuations;
- voltage unbalance;
- power-frequency variations;
- induced low-frequency voltages;
- d.c. in a.c. networks;
- d.c. ground circuits.

(2) Radiated low-frequency phenomena:
- magnetic fields (continuous or transient);
- electric fields.

(3) Conducted high-frequency phenomena:
- induced continuous wave (CW) voltage or currents;
- unidirectional transients;
- oscillatory transients.

(4) Radiated high-frequency phenomena:
- magnetic fields;
- electric fields;
- electromagnetic fields;
- continuous waves;
- transients.

(5) Electrostatic discharge phenomena (ESD).

the required attenuation of RFI under the specified conditions of the standard. Special glands have been developed to allow cable entry into enclosures without reducing their immunity. Manufacturers can establish the EMC compliance of their equipment to the standard by submitting it to appropriately equipped test laboratories.

The International Protection (IP) standard, IEC 529, has an equivalent European norm, EN 60529. International Protection numbers have two digits, e.g. IP 67. The first indicates both the protection offered against dangerous parts of the equipment and the protection of the equipment from ingress of solid foreign objects and dust. This is covered by the scale 0 to 6, in which 0 means no protection, while 6 means dustproof and having no access that would allow even a wire of 1 mm diameter to penetrate into the enclosure. The second number designates the degree of protection provided against ingress of water, on the scale 0 to 8 (see Table 2.6). Here again, the integrity of the enclosure can only be maintained if appropriately designed cable entries are employed. Specially equipped laboratories for IP testing of equipment are also available.

Calibration and maintenance

The user should receive from the manufacturer or supplier the information

Table 2.6 International Protection standard: degree of protection against ingress of water.

Degree of protection	Protects against
0	No protection
1	Drops of condensed water
2	Drops of falling liquid
3	Rain
4	Splashing from any direction
5	Water jets from any direction
6	Conditions on ships' decks
7	A limited period of immersion in water at a stated pressure
8	An unlimited period of immersion in water at a stated pressure

needed to ensure the continued, reliable operation of an instrumentation or control system. This includes the following:

- The requirement for calibration and checking, either on- or off-site. No calibration can be relied upon for ever. In this context it is good practice to have available an auxiliary instrument which can be used as a check on the proper functioning of an automatic control system. Defective working of these systems is not always immediately apparent. For example, a good quality mercury-in-glass thermometer can provide a check on the performance of an environmental temperature control system, and at little expense.
- The necessary steps required to maintain sensors and their associated connections in good order. These are often the most vulnerable elements of a system.
- The maintenance requirements of battery and mains power supplies. In the latter case, it may be advisable to employ a power filter to remove surges, spikes and RFI from the incoming supply, or to install an uninterruptible power supply (UPS), providing battery back up as well as the functions of the power filter.
- Details of the type and amount of instruction and training available to the user.
- Details of the availability of service agreements, which can be particularly important in installations such as crop stores, where large amounts of a commodity can be at risk.

Given this information and advice the user should establish routines to ensure as far as possible that equipment is fully functional when it is needed, that spares of vulnerable items are available and that specialist back up is employed for routine servicing as well as emergencies.

Finally, although industrial instrumentation can be designed to withstand extremely adverse environmental conditions its cost rises steeply with the degree of immunity required. In general, equipment for agricultural applications is in the lower regions of the cost spectrum. Therefore, although it may have immunity to IP 65 level, or even IP 67, its exposure to moisture, dust and corrosive materials should be minimised. In particular, ignitable crop dust is a well known fire hazard and even a small spark is capable of causing ignition.

2.5 Further information

There are many general textbooks on instrumentation and control, providing the theory appropriate for college syllabuses. There is also an increasing amount of material available on CD-ROM and on the WorldWideWeb (WWW).

A considerable amount of information on intelligent instrumentation is supplied in software form by major international companies. For example, details of Intelligent Instrumentation's Visual Designer can be found at http://www.instrument.com. Laboratory Technologies (http://www.adeptscience.co.uk in the UK) produce Labtech® Notebook™ and, like others, use Microsoft's Visual Basic programming language to provide a toolbox for building virtual instruments. National Instruments (http://www.natinst.com) produce LabView. All provide facilities for communication via the serial and parallel links listed in this chapter.

Likewise, data processing and image processing packages are available from internationally known companies such as Mathsoft (http://www.mathsoft.com) – developers of Mathcad – and MathWorks Inc (http://www.mathworks.com), whose MATLAB products include fuzzy logic and neural network toolboxes. These and other companies provide toolboxes for signal analysis and control systems analysis, many of which are oriented to instructional and educational use.

The IEEE's New Technology Database is on the Web site http://www.ieee.org, with a section for the Instrumentation and Measurement Society. This contains updates on hardware standards in the field of data acquisition, *inter alia* (http://www.ieee.org/newtech/reports/im/report.html).

Among published work the background to fuzzy logic, neural networks and genetic algorithms, respectively, can be found in the following:

Goldberg, D.E. (1989) Genetic algorithms in search, optimization and machine learning. Addison-Wesley, Reading, Massachusetts.

Rumelhart, D.E., Hinton, G.E. & Williams, R.J. (1986) Learning internal representations by back-propagating errors. *Nature*, **323**, 533–6.

Zadeh, L.A. (1965) Fuzzy sets. *Information and Control*, **8**, 338–535.

The EU directives on EMC and low voltage were published in the *Official Journal of the European Communities* between 1973 and 1992. In the UK the Department of Trade and Industry (DTI) can supply details of centres which provide full details of these directives.

Chapter 3
Agricultural Crop Production

3.1 Introduction

This chapter deals first with the aerial and soil environment which is the foundation of field crop production. Sections on crop establishment, treatment and harvesting follow, before general sections on field machinery control and field data management.

3.2 The crop environment

This section is concerned with meteorology, soil properties, soil water, irrigation and drainage. The instrumentation aspects of these subjects are taken in turn.

Meteorology

Meteorological information and forecasting is routinely available to arable farmers via the telephone, radio and TV, while a PC with a modem or a shortwave radio link can provide advanced weather information directly or indirectly from weather satellites. Nevertheless, there is still a valuable place for on-site weather stations (Fig. 3.1(a)) which can provide up-to-date data on meteorological and soil conditions. These data can be used as aids to management decisions and as inputs to automatic control of water utilisation. Decisions on the timing of tillage, planting, irrigation, crop protection, fertilisation and harvesting can all be put on a sounder footing by reference to a local weather station's current or accumulated records.

Site-specific crop disease forecasting by reference to combinations of weather variables has been an application of on-farm weather stations for more than a decade. The measured variables supply inputs to forecasting models of environments conducive to the development of a particular infection. Simple models have been used to assess disease risk in wheat and barley, oilseed rape, potatoes, peas and hops. For example, the likelihood of onset of potato blight has been assessed with an ADAS model, which generates a three-point disease risk index. The index is determined by the conditions that potato blight requires at least two consecutive days when the minimum temperature is above 9.5°C and in each day

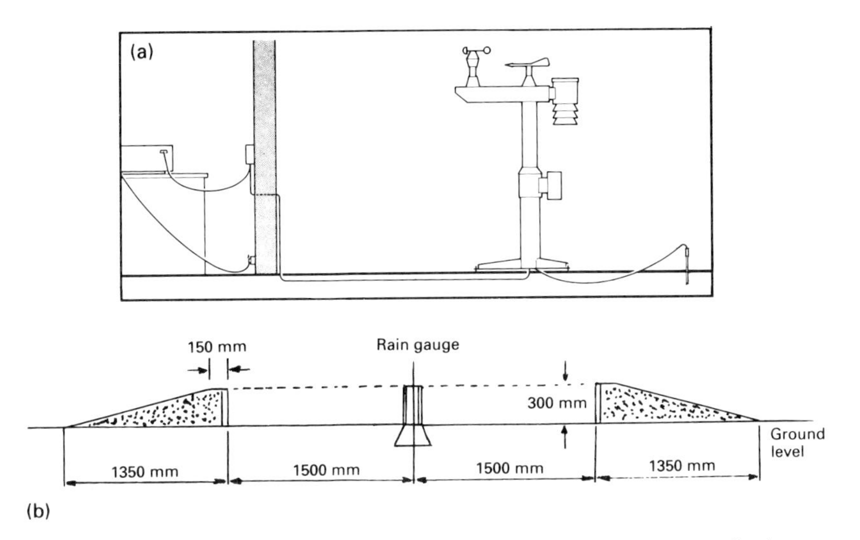

Figure 3.1 (a) Weather station (Vicon); (b) protective bank for an exposed rain-gauge site.

there are more than 10.5 hours when the RH is 90% or higher. Then, if the index is I, the number of hourly observations at which RH $\geqslant$ 90% is n and the minimum temperature between 2100 and 0900 nightly is T, the rules are:

IF $n < 32$ OR $T < 9.5$ THEN $I = 0$
IF $n \geqslant 32$ AND $T \geqslant 9.5$ AND previous $I = 0$, THEN $I = 1$
IF $n \geqslant 32$ AND $T \geqslant 9.5$ AND previous $I \geqslant 1$, THEN $I = 2$

Rather more complex rules are needed for barley brown rust, Rynchosporium and some other diseases, but all are within the compass of simple data processors.

Once a significant risk has been established the meteorological data will determine the appropriate time for safe and effective application of crop sprays.

The above rules were devised by experienced agronomists and they represented an early example of the knowledge-based *expert system* applied at farm level. Similarly, the accumulation of degree days (the product of air temperature and time) has been employed empirically to determine the earliest viable date for fertiliser spreading on grass. Other uses of meteorological data have invoked physical models of environmental processes. Perhaps the best known is the formula developed by H.L. Penman to calculate evapotranspiration from crop cover. This mathematical relationship between daylight hours, solar radiation, air temperature and humidity, and wind velocity has been employed for irrigation scheduling over many years.

Most field operations depend on soil conditions, too. Soil temperature affects the timing of sowing and planting, of course, and it is important at root harvesting. Soil moisture affects the timing of almost all operations throughout the year. Unfortunately, measurement of the latter quantity is not simple and for that reason it is not commonly included in weather station equipment. Methods of measuring soil moisture are reviewed later in this section.

Sensors

The measurements usually made by agricultural weather stations are air temperature and humidity, wind speed and direction, solar radiation and rainfall. Some stations also include soil temperature, barometric pressure and dew sensing (for its connection with grain harvesting, in particular). The main characteristics of most of the sensors used have been described in Chapter 1, section 1.2. Commonly, air and soil temperatures are measured by thermistors, with an accuracy between ±0.1°C and ±0.2°C. The air temperature sensor must be housed in a double-walled, white painted and aspirated radiation screen, which it normally shares with the RH sensor. This arrangement is akin to that in the wet and dry bulb hygrometer shown in Fig. 1.6(b), but the use of a thermistor instead of the more expensive resistance thermometer element precludes the comparison between the dry bulb (air temperature) element and a wet bulb element. Therefore the RH element is normally of the capacitative type. The soil temperature thermistor is sheathed in a spear-shaped probe, usually of stainless steel.

The rotating cup anemometer can be expected to operate up to a maximum wind speed of at least 30 m/s with linearity and accuracy of about 2%, above a threshold value of about 0.5 m/s. Its rate of rotation is usually measured by counting the pulses produced by a reed switch, which is operated by a magnet on the rotating shaft. The wind vane's output is normally produced by a circular potentiometer with an almost 360° winding and a slider which rotates with the shaft.

The solar energy sensor is likely to be a silicon photodiode, facing vertically upwards and with a corrective optical filter which provides a moderately uniform response over the wavelength range 400 to 1000 nm. This combination of the visible and near-infra-red regions of the electromagnetic spectrum (Table 1.4) covers the peak of the sun's radiation energy spectrum and its photosynthetically active region (PAR) from 400 to 700 nm. Ideally, the sensor should be able to receive radiation over the whole hemisphere from 360° in the azimuthal plane to 90° in the vertical plane. Its response should not be affected by the azimuthal angle of incidence of the radiation and its response to radiation at an angle to the vertical should vary according to the cosine of that angle (cosine response). It should also have a low dependence on the ambient temperature. The photodiode can have a satisfactory linearity of ±1%, similar azimuthal accuracy over 360° and a low temperature dependence at –0.1% per °C. However, its cosine response is unlikely to equal that of the solarimeter (see section 4.4), which is a working standard for this measurement. This may be no better than ±5%: even so, that is acceptable for most field applications.

There are several standardised forms of rain gauge, all with the common feature that they collect rainfall over a sharply defined circular area then funnel the water into a lower collector. The collector can be a graduated cylinder with a narrow inlet or neck, to minimise losses by evaporation, but in many cases it is of the tipping bucket type, which lends itself to automatic data collection. This has a divided collector, pivoted at its base, so that only one half can receive precipitation at any time. When the preset amount of water has been collected the mechanism tips, emptying the full container and bringing the empty one into position as the collector. Each tip of the mechanism registers a count. The unit needs to be carefully levelled but it is otherwise robust and can register increments in rainfall as low as 0.2 mm per tip, given an upper collector of diameter 254 mm. Loss of collected water by splash-out is minimised by the design of the upper collector, which has a cylindrical section above its funnel.

Dew sensors are normally of a simple type in which the resistance or capacitance of a flat, horizontal element changes markedly when surface moisture is deposited on them.

Positioning

Placement of the sensors is important because it determines the utility of the data gathered and its compatibility with wider meteorological records. In fact, an approved meteorological station has to conform to very specific requirements on siting, ground cover and positioning of the sensors, for these reasons. Such stringent requirements are neither necessary nor practical in most farm applications but some guidelines should be followed.

First, the weather station should be sited in the open and on level ground, as clear of buildings, walls and trees as possible (50 m or more), and firmly attached to the ground. If the assembly consists of a mast or posts carrying a cross-beam with sensors mounted on it (as many commercial units are) then the beam must be horizontal. If the data gathering and display unit is remotely sited, rather than mounted on the mast, then the connecting cable must be well anchored and protected against traffic and other hazards. This arrangement implies the use of a convenient room or hut at a range of 50 m or more from the station.

The positions of most of the sensors are normally fixed by the supplier, to conform to normal practice. The wind sensors are commonly set at 2 m above ground level, while the air temperature and RH sensors are at 1.5 m. Some units have additional air temperature and RH sensors at 0.2 m height, i.e. nearer to crop level. Rain gauges are sometimes mounted on cross-beams, alongside the wind sensors. However, for greater conformity with strict practice they should be firmly fixed at ground level, with the horizontal rim of the primary collector 300 mm above that level. In very windy sites a surrounding, protective wall is recommended (see Fig. 3.1(b)). Partly embedding the gauge in the ground ensures its stability.

Data processing and transfer

Collection and processing of data from the sensors is performed by an environmentally protected data logger on the mast in many cases. In others the data are transferred by a cable link to an instrument room or office, as already mentioned. From there transfer to a PC is via a RS232C link. Some field units can transmit their data directly to a PC by UHF link, over line-of-sight distances up to 5 km, and others download the data into a PCMCIA card for an office PC with a suitable card reader. Once in the PC, the data can be transferred to other meteorological sites and data centres, via a modem.

Data from the sensors are sampled at intervals that vary from vendor to vendor. Some vendors provide different sampling rates for different sensors. In general, sampling rates are measured in minutes rather than seconds or hours. Most vendors offer maximum/minimum, averaged and total readings of sensor outputs over a given period, which may be daily or over an interval chosen by the user. All provide numerical displays and some also provide graphics.

Maintenance

Although solar power packs are available for field units most are powered by batteries that need replacement or recharge at intervals (usually of the order of 1 year). Apart from that, the main consideration must be the state of the sensors. Cleanliness is particularly important for radiation and dew sensors. The anemometer and wind vane must be undamaged and move freely.

Soil and soil water

Measurement of soil properties in relation to their influence on plant growth has received massive attention over the years, as its importance merits. The complex interactions of the soil matrix with water and other essential root requirements, coupled with concern over soil erosion, soil compaction and pollution of watercourses by heavily loaded water run-off, ensures that research on soils and soil water continues, world wide. To meet research needs, a wide range of specialist instrumentation has been developed, some of which is also in the hands of farm advisory services. However, instrumentation in the hands of farmers is harder to find, although the soil temperature probe has just been referred to. The contents of this subsection are therefore mainly about equipment which is generally employed by specialists, together with developments which appear to have potential for farm use but have not yet been taken to that stage.

Soil bulk density

The use of gamma radiation (see Table 1.4) for soil density determination was established in the 1950s and has become a common tool for research and in some advisory work. The relationship is complex because the gamma rays interact with the soil in two ways:

- Single or multiple scattering of the gamma ray photons by the electrons in the soil, whose numbers bear a straightforward relationship to the soil's macroscopic density. This is known as the Compton effect (A.H. Compton, 1922).

- Attenuation by absorption of photons, dependent on the relative proportion of the atomic elements in the soil and their mass absorption coefficients (μ) at the energy level of the gamma rays.

Together they yield a relationship between the attenuated radiation received by the detector, *A*, and the soil density, ρ, of the form

$$A = \frac{1 + k(\mu\rho d)^n}{e^{\mu\rho d}}$$

where *d* is the distance between the source and the detector, and k and n are constants for the specific application.

Both interactions increase the soil density but at different rates. At low densities scattering dominates and the detector's output increases with the density. At a point dependent on the distance *d* absorption becomes the dominant effect and the output falls progressively with the density. With typical dry soil densities in the range 1 to 2 t/m^3 the calibration can be in the upward part of the curve, the downward part, or both, as *d* varies over 50 to 150 mm. These characteristics have important implications for thc design of the measuring equipment, which is of two general types.

The backscatter meter employs a source and a detector separated by a radiation screen, commonly of lead. The detector then receives a preponderance of scattered radiation. The source and the detector can be coplanar, as they are in sensor heads that are placed on top of the soil, or collinear in a probe that can be pushed into the ground. A disadvantage of this method is that the scattered photons lose energy at each interaction with an electron and the amount lost depends on the scattering angle. Therefore the detector receives radiation with a range of energies, instead of the original monochromatic radiation from the source. In addition, the sampled volume of soil is not well defined. The transmission meter has the source and the detector in two separate probes, as shown in Fig. 3.2(a). This provides a well collimated system which limits the contribution made by scattered photons to those scattered less than 90°, thereby removing those which have suffered significantly large energy loss. Overall, therefore, the sampled volume is much more closely defined and the sensor's output less fuzzy, which is advantageous for measurement of soil density profiles. The surface sensor is of advantage for area surveys.

In both types the radiation source of choice is caesium 137. This radioisotope produces a monochromatic stream of gamma photons with energies of 0.66 MeV (from Table 1.4, 1 MeV = 1.6×10^{-13} joules of energy approximately), which provides for the generation of a high proportion of Compton scattering. In addition caesium 137 has a long half-life of 30 years, therefore the intensity of the source is essentially stable throughout the duration of most experiments.

The detector is normally a scintillation counter (see section 1.8). This has an advantage when it is used with the backscatter meter, since it is possible to filter out the effects of photons with considerably reduced energy by setting an

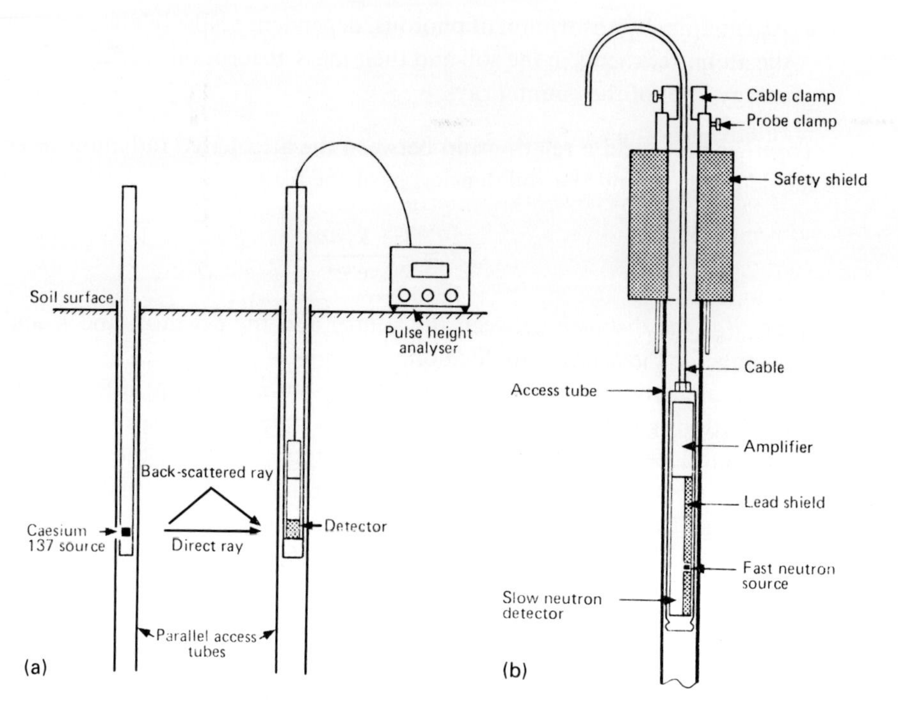

Figure 3.2 (a) Soil density meter (gamma ray attenuation); (b) soil moisture meter (neutron moderation).

appropriate pulse height threshold. Calibration of the output is done with samples of known density.

It should be noted that the moisture content of the soil affects the measurement, since hydrogen and oxygen atoms contribute to both the Compton scattering and the absorption. Therefore gamma ray transmission can provide the basis for soil moisture determination. However, this requires some means to compensate for any swelling or shrinking of the soil that accompanies changes in moisture content. To that end a dual source system has been devised, using americium 241 as the second source. Americium 241 (432 year half-life) emits lower energy radiation at 60 keV approximately, which is close to a wavelength of 2×10^{-11} m. [*Notes* The relationship between energy in eV and wavelength, λ, in metres is provided by Table 1.4. Half-life is the period over which the rate of photon emission diminishes by 50%.] At this energy level the attenuation by water is considerably lower than that of soil. This makes it possible to separate the attenuation due to the soil from that due to the water, by processing the two sets of output signals (of different pulse heights) generated by the scintillation counter.

Soil water

The subject of soil water and its availability to plants is the substance of numerous textbooks, course notes and published papers. It has also given rise to extensive literature on methods of soil moisture determination which can only be treated selectively here. Most of these methods are largely employed in research and advisory work but increasing concern over the supply of fresh water, coupled with advancing technology, may lead to industry adoption of some of them, as indicated in the following text.

Broadly, the instrumentation is concerned with soil water in two zones – the saturated zone and the unsaturated zone in which there is a range of moisture contents between saturation (field capacity) and a permanent wilting point. Each soil has a characteristic relationship between soil moisture and soil suction (matric suction), the latter being a measure of the pressure deficit in the soil relative to that of a free water surface at the same elevation. This quantity has an associated unit, the pF, defined by $pF = \log_{10}h$, where h is in cm water. Table 3.1 shows the relationship between three broad soil conditions and soil suction expressed in pF, cm of water and pressure.

Table 3.1 Soil suction.

Soil condition	pF	cm water	Pressure (bar)
	7	10^7	10000
Dry	6	10^6	1000
	5	10^5	100
	4.5	32×10^4	36
Moist	4	10^4	10
	3	10^3	1
Wet	2.7	500	0.5
	2	10^2	0.1
	1	10	0.01
	0	1	0.001

Three devices for long-term installation in the soil have been employed for many years. The piezometer is used in saturated zones for measuring positive water pressure at different depths. This includes a form in which water passing through a porous tip displaces the diaphragm of a pressure sensor, so providing the means for remote reading and recording. The tensiometer measures matric suction up to about pF 2.9 (0.8 bar), while the electrical resistance soil moisture block takes this measurement down to the wilting point. The tensiometer and the resistance block are both fairly simple in design and are inexpensive devices, but with different spheres of application at present.

The tensiometer is essentially a metal or water-impervious plastic tube terminating in a porous ceramic pot at its base and with a vacuum gauge at its top. The whole system is filled with pure air-free water. When the pot is in close contact with unsaturated soil, water leaves the instrument until the internal and external suctions are balanced. The gauge then reads the matric suction. This device has a long established commercial market, particularly in the horticultural sector, where precise irrigation scheduling is especially important. When the vacuum gauge is replaced or augmented by an electrical vacuum sensor the tensiometer can provide an on-line input to an automatic irrigation system.

The size of the largest pores and the thickness of the ceramic element control the speed of response of the tensiometer. Larger pores provide faster response, but they also shorten the working range, since they reduce the level of suction at which the instrument can become inaccurate as a result of air breaking into the pore space. Response time is also shorter with a thinner element, but here the mechanical strength of the ceramic sets a limit. To achieve its maximum range of 0.8 bar the pore size must not exceed 1.5 μm, which results in a time constant of the order of minutes. Larger pore size can reduce this to seconds, which is more appropriate for automatic irrigation. However, under drier conditions the system's response time may be determined by the soil's hydraulic conductivity, which governs the rate of water transport to the ceramic element.

In use, the element is placed at a depth in the soil which depends on the rooting characteristic of the crop. In deep-rooted crops measurements may be taken at two depths, divided equally between the upper and lower root zones, while in shallow-rooted crops only the lower zone is normally used. The total number of units installed in a given area depends upon preliminary assessments of the spatial variation in matric suction.

The system must be kept air-free, to maintain the instrument's response, and topped up as necessary with air-free distilled water or with the solution supplied by the manufacturer. Given proper maintenance and shielded against extreme temperatures, the tensiometer can be a very reliable instrument, reading suction directly, without need for calibration. Its accuracy depends on that of the vacuum gauge or sensor.

The original development of soil electrical resistance blocks is generally ascribed to G.J. Bouyoucos (1939). They are made in a variety of configurations but their common features are two non-corrodible metal electrodes embedded in a block of porous insulating material. The electrodes are either parallel or concentric, in rod or mesh form, and the total dimensions of the block are typically 50 mm length and 600 mm^2 cross-section.

When the block is buried in soil an exchange of water takes place between the soil and porous material, until the latter reaches an equilibrium moisture content dependent on the relative suction characteristics of the soil and the block. The electrical resistance measured between the electrodes is related (non-linearly) to the moisture content of the porous block.

The function of the insulating material is to minimise changes of electrical contact at the electrode interfaces, and to fill all the space that contributes significantly to the measurement of resistance. The latter requirement means that the electric field between the electrodes is largely contained within the insulator volume. Its pore volume and size distribution determine the useful range of the device to a considerable extent.

The accuracy of measurement with these blocks is affected by hysteresis in their absorption–desorption characteristic, and by variations in the electrical conductivity of the soil water which follow from variations in solute strength and soil temperature. The effect of dissolved salts on the measurement represents the main limitation of the resistance block method.

The electrode material is usually stainless steel, while the insulating material may be nylon or fibreglass fabric, or plaster of paris (gypsum). A thermistor may be incorporated for temperature measurement.

The plaster unit is inherently less durable than the other types and its characteristics are less stable, due to the solubility of the material. However, the presence of dissolved calcium sulphate in the pores of the block helps to reduce the effect of soluble salts in the soil water (buffering action) and the addition of nylon resin during manufacture has been shown to increase the durability considerably, although at some loss of sensitivity.

A moderate degree of uniformity between blocks of the same type can be obtained by careful manufacture. Their useful limit at the wet end of the suction range varies from type to type, and may lie anywhere between 0.5 and 2 bar.

Accuracy of measurement with these blocks is not high, i.e. less than 5% unless careful calibration is undertaken. Their time constants vary according to type, but they can be many minutes.

Measurement of resistance is always carried out under a.c. conditions, to avoid polarisation effects. Frequencies in the low kHz range are used. The meter requires a wide-range bridge circuit (100 Ω to 1 MΩ at least) and additional capacitance balance facilities, due to the large capacitance component of the block's impedance. Portable, weatherproof meters are available for the measurement.

A single meter can be used in turn with many *in situ* blocks to provide a low-cost means for regular sampling of soil water conditions over a wide area. Nonetheless, these devices remain more of a tool for experimenters than for commercial farmers.

Electronic instruments

Here a variety of more complex methods currently in use for soil water determination are reviewed briefly.

Soil moisture determination by gamma-ray absorption has been mentioned, but the neutron moderation method (developed in the 1950s) is more widely used. Figure 3.2(b) shows one form of the instrument, which bears a close resemblance to a back-scatter gamma ray probe. In this case the source is an alpha-ray emitter (usually americium) mixed with beryllium. The alpha particle is the helium nucleus, comprising two protons (positively charged) and two neutrons (electrically neutral). It is produced in the radioactive decay process of americium and other elements and, like X-rays, gamma rays and neutrons, requires close attention to safety precautions, although its penetrating power in solid material is limited.

The americium/beryllium mixture generates carbon 12 and an energetic neutron (5.65 MeV) at each alpha particle–beryllium 9 interaction. The emergent neutrons then interact with surrounding matter, being absorbed or scattered, in rather the same way as gamma rays. After multiple scattering neutrons are finally reduced to thermal neutrons with a low energy of about 0.025 MeV. As stated in section 1.15, the process of neutron *thermalisation*, or *moderation*, is greatly enhanced as the atomic mass of surrounding atoms diminishes, and the fractional energy loss per collision with the hydrogen atom is considerably greater than that for any other nucleus. Therefore the relationship between water content in a material and the production of thermal neutrons can be close.

The detector is a proportional counter – a tube filled with boron trifluoride gas in which the thermal neutron reacts with the boron atom to create lithium and a 2.8 MeV alpha particle. The tube has electrodes to which a high voltage (around 1 kV) is applied. The ionisation of the gas created by the alpha particles creates a brief pulse of current through the tube per alpha particle generated, and an output pulse ensues. The water content of the surrounding matter is deduced from the resulting pulse rate, via a calibration procedure.

Figure 3.2(b) shows that the neutron source can be withdrawn into a safety shield when not in use. The shield also has a calibration function when the source and detector are in the parked position, because it employs a massive amount of polythene, or a similar material with a high hydrogen content. Nevertheless, variations in the density and chemical composition of soil require some reference to samples of soil (or soils) with known moisture contents, either directly or via established calibration curves for the soil type.

The radius of influence of the neutron probe, R, can be calculated by reference to a relationship due to C.H.M. Van Bavel:

$$R = 15\left(\frac{100}{M_v}\right)^{\frac{1}{3}}$$

where R is in cm and M_v is the percentage moisture content by volume. This yields a range of 15 cm ($M_v = 100\%$) to 70 cm ($M_v = 1\%$). The range in dry soil is about 30 cm. However, in practice the sphere of influence has been more accurately described as an ill-defined ellipsoid.

Section 1.15 also introduced the basic concepts of nuclear magnetic resonance (NMR) for moisture determination, based on its sensitivity to the hydrogen content of materials. In this case, a sample (or sampled region) of a hydrogen-rich material is placed in a combined magnetic and electric field. Tuning of this combination to a resonant frequency of the hydrogen nuclei (protons) causes them to absorb energy from the RF generator in proportion to their abundance. The technique has attracted attention as a research tool for nearly 40 years. It has been shown that it is comparable with the gamma and neutron techniques – and without their hazards – given calibration for soils of different types, structures and chemical composition. However, the calibration curve shows three distinct sectors, corresponding to soil below permanent wilting point, an intermediate sector and near saturation. Despite these complexities, in the 1980s research workers developed a tractor-mounted, soil contacting version that was capable of measuring soil water at seed planting depth down to about 60 mm.

Major advances in the production of compact, high-powered magnets, coupled with those in computer-based data processing, have made this type of equipment less costly and cumbersome than it once was. Therefore, further development in the field machinery context is possible.

The more refined magnetic resonance imaging technique (MRI) is based on the use of auxiliary coils which create a linear magnetic field gradient in the sample. This makes it possible to create proton resonance in successive *slices* of the sample and so to build up a three-dimensional pattern of water distribution in it, which can be displayed in simulated three-dimensional form on a computer screen (see section 5.4).

Another set of techniques depends on the dielectric constant of liquid water. Dielectric constant, or relative permeability, is a measure of the polarisation of captive charges in a material when it is placed in an electric field. Water has a relative permeability of about at 80 at frequencies below 3 GHz, whereas the figure for gases is almost exactly 1 and for solids (including ice) it is mostly in the region 2 to 6. Therefore measurement of this property of non-conducting materials is widely employed for determination of their moisture content.

The first of these techniques is the measurement of capacitance. A volume of the moist material is placed between two electrodes – commonly in the form of parallel plates or concentric cylinders – to which an alternating voltage is applied. The measured capacitance of the assembly is then related to the moisture content of the enclosed material. The classic a.c. impedance bridge circuit – akin to Fig. 1.4 but energised at a radio frequency – is probably the commonest form of capacitance meter, but another well known technique is to make the sample cell part of an oscillating circuit. The frequency of the oscillator then changes with the sample's moisture content.

It is necessary to perform the measurement at RF because at lower frequencies the electrical conductance of the sample can have a marked effect on the result. Salts in the soil water will make it conductive, apart from anything else in the material. This point is illustrated in Chapter 5. Work done in the 1960s showed that a soil moisture probe, energised at about 30 MHz, produced a moisture calibration curve that was practically independent of soil type above 5% free water content by volume. At lower levels, as with the NMR method, the uncertain influence of water molecules tightly bound to the minerals in the soil matrix can become significant. They have a relative permeability close to that of ice. The measurements were performed with electrodes attached to the tip of the probe, therefore the sensor responded to the moisture in the soil surrounding the electrodes, rather than soil sandwiched between them. In this configuration the measured capacitance is that of the soil within the *fringe field* of the electrodes, which is a small volume of imprecise extent, characterised by diminishing field strength at increasing distance from the probe. That contrasts with the parallel plate or concentric capacity measurement mentioned above, where the influence of material in the fringe field at the edges of these capacitors is something to be minimised.

Later, commercial developments of the probe system have increased the frequency of measurement to over 100 MHz. The capacitance probe's independence of soil type has also led to evaluation of its suitability as an attachment to field machines, for dynamic measurement of soil water in the cultivated zone. Reported results have not discouraged this approach.

Time domain reflectometry (TDR) is another measurement based on relative permeability, and one which has attained increasing prominence over the past decade. Following initial development by J.L. Davis and W.J. Chudobiak (1975), work by G.C. Topp and coworkers (1980) established an empirical relationship between the relative permeability of a soil, κ, and its volumetric moisture content, θ, in the form

$$\kappa = 3.0 + 9.3\theta + 146\theta^2 - 76\theta^3$$

This had proved widely independent of soil type and bulk density, soil temperature and salinity, thereby dispensing with the need for calibration with different soils. Additionally, the soil probes (to be described below) are relatively simple devices. Therefore the method had much to commend it and commercial development followed. The software incorporated the above relationship, which is easily handled by a data processor.

Subsequent research has shown that there are discrepancies in the case of soils of fine texture or with high organic content. In consequence, a more complex relationship has been established through modelling the system by reference to the theory of dielectrics. The relationship includes the relative permeabilities for bound water ($\kappa = 3$ approximately) and soil. Its use requires the measurement of the soil's bulk density (in t/m^3) and its specific surface in (m^2/kg). Both of these quantities can be measured by standard methods.

Essentially, the method and apparatus used are akin to those employed in testing buried electric cables for the presence of short circuits and open circuits. This is done by injecting a subnanosecond electrical pulse into the cable, then measuring the time delay to the arrival of a reflection from any form of discontinuity. The cable acts as a transmission line and the velocity of travel of the electromagnetic wave along it depends on the relative permeability of the surrounding medium. In the case of cable testing the value of κ must be known in order to determine the wave's velocity and hence the distance to the discontinuity. Here the κ is that of the cable's insulation, which is usually well documented. Conversely, in the TDR system the length of the buried transmission-line probe determines the distance to the discontinuity (the change in κ at the end of the probe). Therefore the measured time delay provides the required value of κ, from which the moisture content, θ, can be deduced.

The probes can be hand-held or inserted in the soil and left in place, like electrical resistance soil blocks, then coupled to the TDR transmitter/receiver as required. For major surveys involving a cluster of sensors a SCADA system can be employed. The probes are permanently connected to the TDR unit via a multiplexer which enables the supervisory computer to select channels in turn. Data from the TDR unit are transferred to the computer by an RS232 or similar link for subsequent processing and analysis.

This measurement has been extended to measurement of soil salinity by determination of the soil's electrical conductivity. Measurement of soil water content and conductivity are made at the same time. For this purpose the probe has developed in the form of a trio of spaced electrodes of about 100 mm length. The delayed reflection from the soil beyond the end of the probe bears a complex relationship to the conductivity of the soil in that region but it is possible to establish a useful working relationship by calibrating the probe in solutions of known conductivity and temperature. In order to compensate for variations of soil conductivity with temperature the probes may incorporate temperature sensors.

These TDR systems are robust and non-hazardous, but the cost of the TDR unit has been a deterrent to its more general use. This may change. In suitable soils they make it possible to measure the volumetric water content to within a few per cent standard error over the range 0.1 to $0.4\,m^3/m^3$, irrespective of bulk density variations of $\pm 0.5\,t/m^3$. On available evidence the measurement of soil salinity is less dependable.

Within the 500 MHz to 5 GHz microwave region, ground penetrating radar (GPR) has potential for subsurface surveys of soil water. This technique has been used as a nondestructive search tool for cables and other buried objects and for mapping subsurface formations of many kinds. The transmit/receive antenna is placed on the soil surface and, like the TDR system, the receiver collects pulses reflected from zones of different relative permittivity, thereby locating the depths of those zones. Deep probing at lower MHz frequencies can extend to over 50 m but for detailed measurements in the cultivated zone the upper range from 500 MHz is necessary. In this range wavelengths are of 200 mm or less in typical soils. Ground penetrating radar has been used to estimate the level of the water table down to 12 m in some soils, with a mean accuracy of better than 50 mm. Experiments using two centre frequencies, to eliminate the effects of bulk density variations in the soil, have shown that the method also has potential for non-contacting moisture determination in the plough layer. This could make it suitable for use on a field machine, like the soil-contacting capacitance probe.

At even shorter wavelengths the reflectance of soil in the near infra-red (NIR) region correlates well with its moisture content, although the measurement is only that of surface moisture. The basis of this measurement is long established and has been applied to moisture determination in many other spheres. Water exhibits molecular resonance at a number of wavebands in the NIR band, therefore if light reflected from the surface is filtered in some way, to select the peak of one of those bands, the moisture content is inversely proportional to the intensity of the reflected light at that wavelength. In order to make the measurement meaningful there must be a zero reference, but soil reflectance is complex, due to factors other than water. Therefore the nearest wavelength that is clear of the water absorption band is used as a reference. The relative reflectance at the two wavelengths establishes the moisture content. Commonly the strong absorption band centred on 1940 nm (1.94 μm) is employed, with 1800 nm as the reference wavelength (see Fig. 3.3). These can be selected by interference filters, which transmit only a narrow waveband. Calibrations of clay-loam soils have shown that ±2% standard error is attainable over the volumetric range 0.05 to 0.35 m^3/m^3. This level of accuracy could not be attained in soils with marked swell/shrink characteristics.

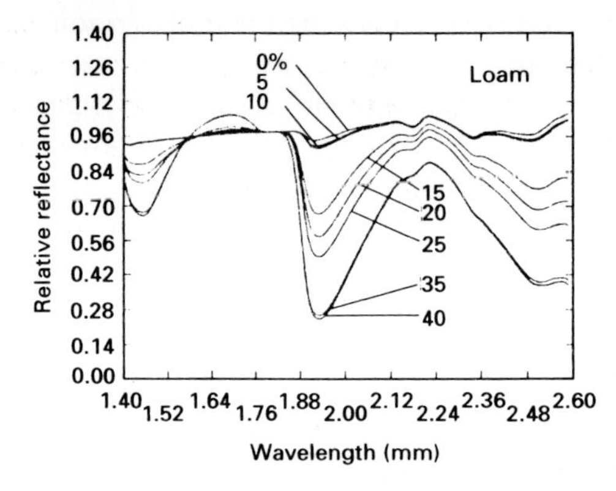

Figure 3.3 Soil moisture measurements, near infra-red (University of North Carolina).

Use of this method for subsurface moisture measurement would depend on the development of a probe with a suitable sensor at its tip. In principle, this requirement could be met by a combination of two LEDs, operating at the above wavelengths, together with a photodetector (see later in this section). To date, such LEDs are not available.

Attempts to use microwave reflectance in the GHz region as a means to monitor subsurface soil moisture have also met with limited success. At the centimetric wavelengths in this region soil irregularities can scatter the radiation strongly, therefore selecting an appropriate angle of incidence is not easy. Furthermore, the radiation's ability to provide a useful measurement at anything deeper than the first few centimetres of the surface layer has not been established.

Evapotranspiration, evaporation and water stress

The loss of water through plant transpiration and evaporation from the soil can be estimated by reference to data from an on-site meteorological station, as indicated at the beginning of this section. Evaporation can be measured more directly if evaporimeters are employed. These are normally in the form of a horizontal porous plate or pan from which free evaporation can take place, and which is fed from a pure water supply in a closed reservoir of some kind. Water lost by evaporation is determined either by the fall in the reservoir's water level or by the amount of water required to maintain that level. The latter method is more suitable for automatic operation, since it employs an on–off control system with a fixed differential, which adds a constant amount of water at each *on* period. The total amount of water evaporated over a period can be measured by counting the periods. The cleanliness of the porous plate and the water supply are crucial to the accuracy and reliability of this system. The relationship between the count and the actual soil water deficit has to be determined by some means, such as starting the count when the soil is known to be at field capacity. The obvious limitation of this method is that the relationship between evaporation and combined evapotranspiration is not a constant one.

The ultimate requirement is to keep plants supplied with sufficient water for their requirements at any stage, but where water is in short supply it may be necessary to allow them to suffer some water stress. This condition is related to the difference between the temperature of the plant vegetation and that of the ambient air. The leaf ambient temperature can be several °C positive or negative, being negative (leaf cooler) if the plants are well supplied with water, and positive if thcy are under stress. The plant temperature can be monitored by an infra-red thermometer, which is one form of the radiation sensors mentioned in section 1.8. This device is calibrated to measure thermodynamic temperature via the relationship that the radiant energy received from a black body is proportional to the fourth power of its temperature in K. Plant leaves have lower emissivity than a black body – i.e. they radiate less energy at the same temperature – therefore the thermometer's reading has to be corrected for this emissivity factor. Overall, this technique seems unlikely to find much application outside very arid regions.

Irrigation and drainage

Reference has been made to closed loop control of irrigation, based on tensiometer measurements, although only in the horticultural context, with its high value production. Many of the foregoing instruments (including the weather station) can supply an on-line input to an automatic irrigation system via a control computer, but their main spheres of application at present appear to be in either the high value sector (which includes golf courses and other leisure areas) or large-scale irrigation schemes in drier regions.

Instrumentation in field irrigation and drainage systems also includes flow meters and pressure gauges in pipelines, and open channel flow meters of the type described in section 1.11. The pipeline instruments operate as the field level devices in SCADA systems for very large piped irrigation schemes. These systems apply optimal control strategies to minimise energy consumption by scheduling pumped water flow through the pipe network in the most economical manner. Flux vector drives (section 2.3) are used in these systems to maintain constant pressures at variable flow rates.

The special irrigation requirements of greenhouses and nurseries are dealt with in Chapter 4.

Other soil measurements

The pH sensor described in section 1.16 (and shown diagrammatically in Fig. 3.4(a)) can be applied to soil measurements by placing it in a mixture of soil and distilled water – typically 20 g of soil with 20 ml of water. The addition of water affects the pH to be measured but it ensures uniform contact with the electrodes. However, for on-farm tests simple and widely available chemical methods are more practical. Equally, the nutrient status of soil is normally evaluated by similar chemical tests or by submission of samples to an analytical laboratory.

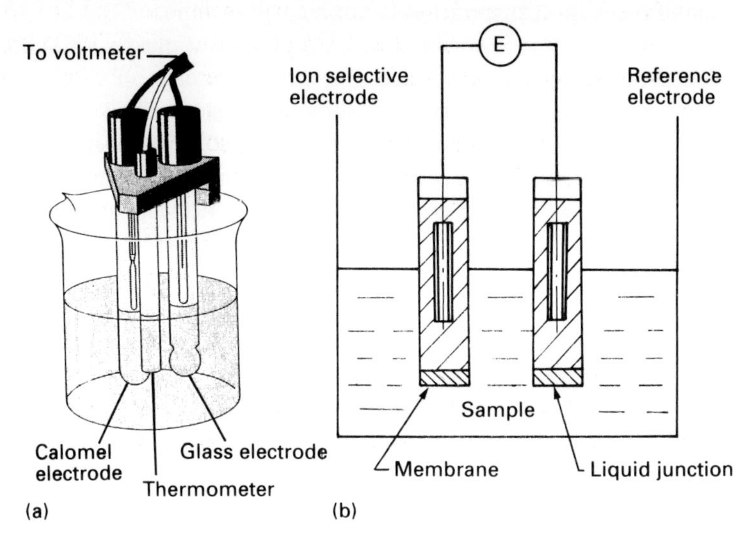

Figure 3.4 (a) pH cell; (b) ion-selective electrode.

Nevertheless there is widespread research activity to identify more methods of physical and chemical analysis of soils that are suitable for field application. These activities are inspired by the technological advances, discussed in section 3.4, which have given rise to the concept of *precision-agriculture* or *site-specific crop management*.

The variability of soil composition and structure has to a large extent defined the level of accuracy attainable by most of the sensors described in this section. Sufficient sampling can provide data for an overall evaluation of the need to irrigate, or the amount of water to apply. Irrigation in general seems an unlikely candidate for site-specific control, like primary cultivation and once-over harvesting. However, placement of seed or seedlings at the optimum depth in relation to soil moisture might repay a site-specific approach. This has been the thinking behind the investigations of the effectiveness of moisture sensors on field machines, outlined in this section. Other aims of research are to monitor soil pH, nutrient status and soil organic matter in a similar way.

Section 1.16 also referred to ion-selective sensors, with specific reference to the modified field-effect transistor known as the ISFET, which responds to the presence of specific ions. Ion-selective electrodes (Fig. 3.4(b)) have been commercially available for over 20 years. ISEs can be used for measurement of nitrate, potassium, sodium and other ions in solution. They were developed from the glass pH electrode, as can be seen from comparison with Fig. 3.4(a). Their limitations are that their performance is dependent to an extent on the type, concentration and activity of other ions in the solution and they need regular calibration as well as flushing to avoid build up of contaminants. Calibration is done with standard solutions. A commercial hand-held nitrate meter based on one of these sensors has been evaluated with soils and found to provide a reliable reading within a few minutes. However, this is rather long for real-time measurements on a field machine. Therefore ISFETs have been evaluated for this purpose.

In contrast to the ISE, ISFETs are small, solid-state devices, with a far more rapid response and the ability to work with a small sample volume. In addition, they have a low-impedance output, which is more convenient for circuit design than the high-impedance pH and ISE sensors. Like the ISE, ISFETs suffer from drift and hysteresis, but this problem can be reduced by rcpetitive injection of the test solution followed by flushing on a short cycle time (2 to 4 s). This provides a zero baseline for each successive measurement. Nitrates have been monitored in this way with promising results.

A further advantage of the ISFET is that multiple elements can be obtained on a single chip, making it possible to sense several different ion species simultaneously.

Soil organic matter (SOM) is another feature of interest in the context of site-specific management, because research has shown that the required amount of some soil-applied herbicides increases with the amount of organic matter present. Experiments in the 1970s established that SOM can be measured with an acceptable degree of accuracy by reference to the spectral reflectance of soil in the red and NIR regions. Since then development has proceeded on two lines.

The simpler method employs an array of LEDs, emitting light at 660 nm wavelength. This combined light source provides the sole illumination for a soil area that is in the field of view of a photodiode sensor. The LED array is designed to minimise the possibility that the sensor will receive light from any preferential direction. Calibration over the range 1 to 5% SOM showed that the sensor needed a separate calibration for individual soil series. After that calibration the correlation between reflectance and SOM could be high (r^2 up to 0.98, in statistical terms). A sensor designed on these lines has been mounted on a tractor toolbar and positioned below the soil surface, where it is clear of plant material and soil surface irregularities, and where the effect of soil moisture variations is minimal. Changes in depth and travel speed have to be allowed for.

A more elaborate method is based on multiple wavelength sensing in the NIR range 1630 to 2640 nm. A portable spectrophotometer was designed for this purpose. This was not intended for on-line monitoring and control but for field mapping to provide data for subsequent site-specific herbicide application, either on its own or in combination with other data in a geographic information system (GIS). The instrument employs a broad-band quartz–halogen light source, a rotating disc filter to provide the required spectrum and a lead sulphide photodetector (one of the original detectors for the NIR region). Sampling and digitisation of the detector's output is synchronised with the spectral scan by an encoder on the filter's shaft. The rotating disc also provides an interpolated reference level of illumination, thereby normalising (standardising) each measurement. A microprocessor performs the data analysis and a complete reflectance spectrum is available in less than 10 s. Laboratory calibration against 30 soils of known SOM content, coupled with statistical treatment to reduce the variability due to the influence of moisture on the reading, produced a calibration that could be used with all the soils between field capacity and permanent wilting point. Conversely, use of the same statistical treatment (partial least squares regression) to minimise the effect of SOM produced a prediction of volumetric soil moisture percentage for the same soils that was more accurate than the prediction of SOM.

Water monitoring

Water run off from agricultural land into watercourses inevitably links agriculture with monitoring of the quality of river water. River water analysis involves monitoring stations at critical sites, such as those near the intake of a potable water plant. These stations employ some of the sensors mentioned in this section. Electrical conductivity, pH and nitrates are among the frequently monitored parameters. Nitrate levels are determined by ISEs. Fouling of some sensors can be a serious problem, requiring ultrafiltration techniques which separate water and dissolved salts from suspended solids and colloidal material over 0.02 μm in particle size. Regular servicing of the equipment is essential, too. However, this does not cover current requirements for water purity, which include maximum levels of herbicides in water supplies.

In that context, a European Union directive has set the collective safe level of mixtures of herbicides in natural water at 0.5 μg/l, so stimulating a pan-European research project to develop a suitable sensor for widespread use. In turn, the project led to the development of a solid-state sensor capable of monitoring simazine and its metabolites down to that level. It is classified as an optoelectrochemical sensor since it operates with 633 nm light from a helium–neon laser. This passes through a waveguide, which is akin to a transmission line, but at the above wavelength can be only a few mm wide and deep. Superimposed on the waveguide is an insulating *buffer* layer covered with a gold layer, typically 50 nm thick. The waveguide is embedded in a silicate glass substrate, machined by photolithography, and the complete unit is mounted in a flow cell. The presence of herbicide in the water affects a chemically specific adsorbed layer on the gold surface, thereby altering the output of the laser radiation from the waveguide, via a process known as surface plasmon resonance. Like the ISFET, this sensor needs to be reset after each measurement and this can be done by flushing it with a weak acidic solution.

More years of research may bring this device within range of the EU's current directive for the safe level of a single pesticide compound (1 μg/l) and the ability to manufacture a multichannel, multicompound detector that can be used outside a specialist laboratory. Nevertheless, chemical and biochemical sensor technology is now the most dynamic area of sensor development, and it can be expected to have a major impact on monitoring and control in the agricultural sector.

3.3 Crop production

This section covers the wide range of field activities from cultivations to harvesting, and the field machines that perform these operations.

Machine monitoring and control

Since the 1980s, the introduction of electronic equipment for monitoring and control of field machinery has advanced from the occasional *bolt-on* accessory to the integrated systems that are available on many production machines and from most manufacturers. This has been brought about by a combination of four factors:

- The need for more and timely information in the interests of more efficient operation.
- The widespread influence of electronics in industrial, social and domestic life.
- Increasing user confidence in the ability of electronic equipment to function reliably in the agricultural environment.
- The increasing user friendliness of that equipment, resulting from improvements in both hardware and software – but especially the latter.

Tractors

The tractor's performance, general facilities, shape and degree of operator comfort and convenience have all improved substantially over the above period. The driver has more options to match machine settings to the immediate requirement of the field task, backed by automatic control systems and in-cab information to supplement or complement direct observation of the task.

Electronic control is gradually becoming established at the heart of the tractor – its diesel engine – following developments designed for use in trucks and buses. One reason for this is that precise field work requires closely controlled engine and pto speeds, maintained under the conditions of variable loading that are characteristic of field operations. Another is the need for precise timing of fuel injection in order to meet impending legislative limits for exhaust emissions by self-propelled machines, on or off the road.

The traditional mechanical governor cannot match the performance of electronic governors for engine speed control, nor can it match the latter's facilities for fine tuning of the engine, nor its long-term freedom from wear. Figure 3.5(a) outlines one form of the electronic control unit (ECU). The demand is provided by the potentiometer actuated by the foot throttle, and the response is measured by the engine speed sensor. The digital output of the latter is converted to an analogue signal for comparison with the former and the error (difference) is submitted to a classical PI or PID controller (see section 2.3). That adjusts the fuel pump in the direction required to achieve or maintain the selected speed. Additionally, the controller must limit the engine speed to its rated maximum,

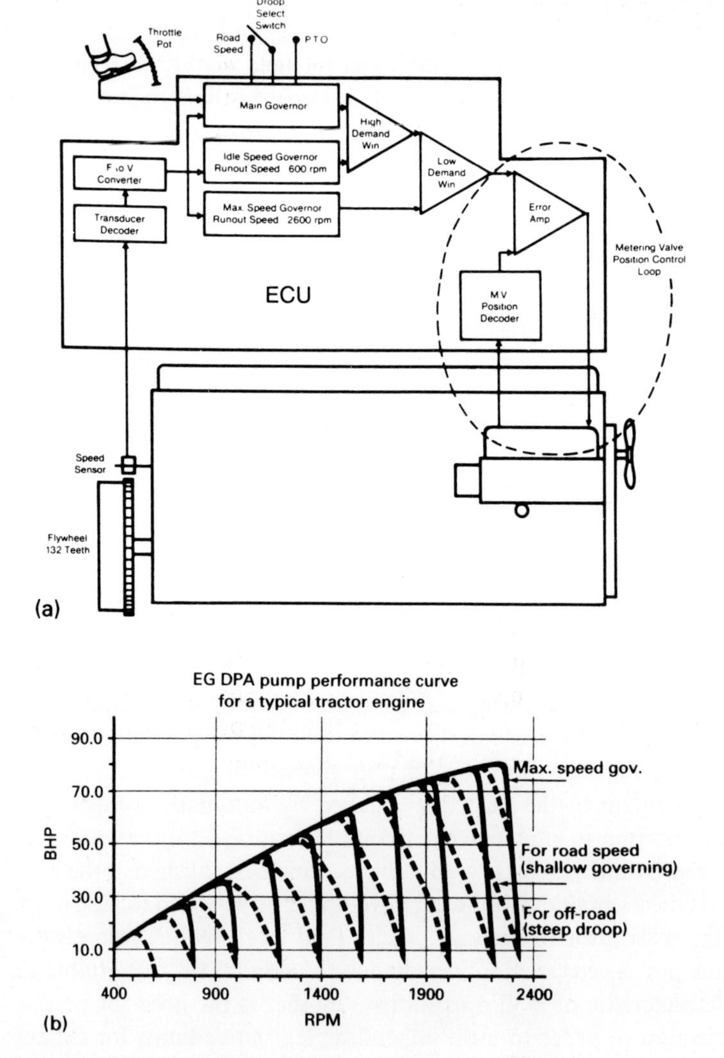

Figure 3.5 (a) Electronic governor for a diesel engine; (b) governor performance (Lucas).

regardless of the throttle input, and at the other extreme it must ensure that the rpm do not fall below the prescribed idle speed. Safety trips must stop the engine if its speed exceeds the rated maximum by a preset amount, or if the speed sensor's output signal is absent.

Another feature is the availability of two control settings – one for field work and the other for road work. In the field the governor's ability to control engine and pto speeds within close limits at any throttle setting, irrespective of loads between zero and maximum, is illustrated in Fig. 3.5(b). Numerically, the closeness of control is expressed as the governor's *droop*, defined as the percentage change in speed between the above load limits. Electronic governors can hold this value to better than 5%. However, to achieve that performance under dynamic conditions the control unit has to respond as quickly as possible to load changes, without incurring the overshoot depicted in Fig. 1.3(c). With the comparatively steady throttle settings employed in field work that is practicable but during manoeuvring or road work tight control can be too responsive to the attendant throttle changes, making smooth operation difficult. Therefore a larger droop needs to be selected for those circumstances. That requires only a simple switch, as shown in Fig. 3.5(a). In fact, with some systems, the close control mode acts as a cruise control, the required speed being set with a keypad.

Traditional injection pumps have been adapted to this form of electronic control by fitting an electrical solenoid to the metering valve. The ECU continuously calculates the position of the valve that will deliver the fuel flow required, by reference to stored data on its characteristics. At the same time a position sensor on the actuator feeds back the valve's actual setting to the ECU. Closed loop control then drives the actuator until the required position is attained.

Timing of fuel injection can also be improved by electronic control, through ECU regulation of the cam-operated pump-advance mechanism on the engine. This has been done in the same way as speed control – i.e. the position of the cam is detected by a sensor which feeds back that information to the ECU. The latter then refers to stored data in its memory and controls an actuator which adjusts the cam to the required position. Another method is based on sensing the lift of the needle in one injector, as the timing reference. Then the ECU adjusts the cam setting until injection begins at the required point relative to top dead centre. This timing can be altered by the ECU according to engine speed and load.

A different approach is required for engines employing unit injectors, i.e.. with the injector at each cylinder operating as its own metering device. In this case the engine's crankshaft drives a relatively low pressure feed pump, together with a camshaft which drives a plunger in each unit. When it is driven by its cam the plunger closes off the fuel feed which otherwise passes through the unit and out via a spill valve into the spill line. The continuing movement of the plunger drives trapped fuel out through the spill valve until an electrical actuator controlled by the ECU closes that valve for a precise time. During that time the small quantity of fuel in the injection chamber at the base of the unit is compressed by the plunger to high pressure (of the order of 1500 bar). The injection nozzle's valve lifts from its seating and injection takes place. As soon as the spill valve is released by the actuator injection stops and the nozzle valve is reseated. In this way the timing of the start of injection and its duration can be controlled to almost microseccond precision. The duration depends on the instantaneous demand calculated by the ECU and in principle it could vary from the zero load demand to the full load demand in successive cycles, so the system is highly responsive. In comparison with Fig. 3.5(a) it requires the crankshaft system instead of the metering pump, and feedback to the ECU from an angular rotation sensor on the camshaft, to provide the necessary synchronisation.

Apart from its control functions, the ECU can provide a range of diagnostic features, especially when additional sensors, such as coolant temperature sensors, are employed. Finally, it should be noted that electronic governors of this type have at least equal value on combine harvesters.

Electronic control has also become established on the next major element of the tractor – its transmission – mainly in the form of microprocessor-based coordination of the control functions associated with powershift stepped ratio transmissions. In this case the operator initiates the shift; thereafter the microprocessor supervises the operation of internal clutches to provide smooth changes by reference to engine speed, forward speed and turbocharger boost pressure, *inter alia*. However, if the operator chooses a shift that could result in damage to the transmissions, the microprocessor will not sanction it. In addition to essential diagnostics such as this, fail-safe features are provided.

Automatic power shift has been added to this facility. The ECU takes data from the engine management system and selects the appropriate gear for current requirements, within limits for forward and reverse gear selected by the operator. These, together with a selected starting ratio, are programmed via a joystick control. The operator thereafter looks after the steering. The clutch is only needed for inching and stopping, while headland turns only involve the joystick. All of these features add to the efficiency and precision of field work, and reduce operator fatigue.

One manufacturer has also introduced a digital rotation sensor at the pto drive. Changes in the phase difference between its output and that of a digital sensor at the crankshaft provide a measure of the torque being transmitted. The ECU automatically adjusts the injection pump to a higher power setting when the power transmitted to the implement exceeds a preset level.

The third major element is the implement control system, preponderantly applied through the three-point linkage devised by H.G. Ferguson (1925). For many years this has been the preserve of pure hydraulic control although an electrohydraulic version was produced by Robert Bosch in 1978. A 1980s form of that system is shown in Fig. 3.6.

The introduction of closed, *quiet* cabs and general improvements in the ergonomics design of their seating and controls have helped to promote the use of these more costly systems, in addition to the improvements in speed, precision

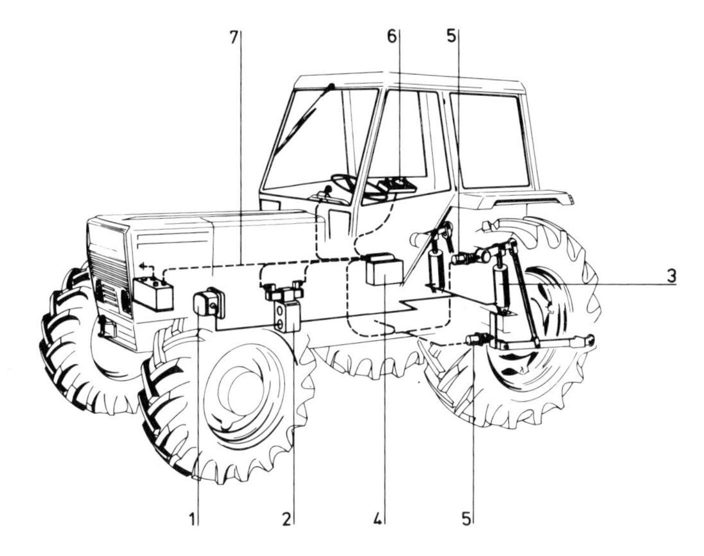

Figure 3.6 Three-point linkage control; 1, hydraulic pump; 2, hydraulic control valve; 3, hydraulic cylinder; 4, ECU; 5, lift shaft and draught sensors; 6, operating panel; 7, cable harness (Bosch).

and flexibility that they offer. In-cab electrical and electronic controls and displays are easier to position for the convenience of the operator and to interconnect with external sensors and actuators, without compromising the cab's other ergonomics features. In operation, electrohydraulic systems do not suffer from the friction, wear, hysteresis and elasticity of the mechanical systems. They use lower-link sensing of draught, now normally employing load pins in the coupling between these links and the tractor chassis (lower 5 in Fig. 3.6). These links are a form of load cell (see section 1.14). The upper link sensor is commonly a linear displacement sensor actuated by a rotating cam on the rotating lift shaft (upper 5 in Fig. 3.6). This provides position sensing when the linkage is lifted.

The tractor's engine transmission and implement control systems are now seen as interacting elements of the integrated tractor/implement monitoring and control systems that are just emerging. However, before moving to this topic several further subsystems will be dealt with.

Ground speed and wheel slip

In order to obtain maximum uniformity in seeding, planting, fertiliser distribution and spray deposition it is necessary to know the actual forward speed, or ground speed, of a tractor. Drive wheels will slip under field conditions, even with light loads. Also, in heavy draught operations excessive wheel slip will create needless tyre wear and excessive fuel consumption. It will reduce work rate and it can seriously affect soil structure. Therefore ground speed monitoring is important for two reasons.

A simple way to determine actual ground speed on 2 WD tractors is to measure the speed of rotation of one of the undriven front wheels. In order to convert this to linear speed the wheel is assumed to roll over the soil surface without slip, skid, yaw or bounce, and to maintain a constant circumferential length at all times. This is an unattainable ideal, of course, but for many purposes a simple system, such as that shown diagrammatically in Fig. 3.7(a), is adequate. The measurement is usually made by fixing magnets to the undriven wheel and mounting a magnetic sensor on the axle support. The sensor produces a digital pulse output each time a magnet moves past it. Therefore, the pulse rate is an indicator of the wheel's rotational speed and hence, on the above assumptions, of ground speed. The distance between the magnets and the sensor is not critical and the whole assembly is environmentally robust. The associated meter must first be programmed with the wheel's effective circumference. A simple adjustment can be provided for this; then the display reads ground speed directly. Calibration can be checked by timing a measured straight run (50 m minimum) over a flat, non-slip surface, at constant speed. The longer the course, the better, since uncertainties in timing the start and end of the run will be of less consequence. This method is not adaptable to 4 WD tractors, though, nor does it always have the accuracy and speed of response for more precise field operations. The requirement in these cases can be met by the Doppler speed meter, powered by the tractor's battery, although at considerably greater cost.

As stated in section 1.10 (and illustrated in Fig. 1.9) the Doppler effect is the apparent change in frequency of radiation that results when a transmitter (or reflector) and a receiver are in relative motion. In this application the radiation is electromagnetic and its frequency is in the microwave (GHz) band. A Doppler meter must operate at an authorised frequency and power output in this band. Also since radar speed meter equipment is licensable in the UK under the

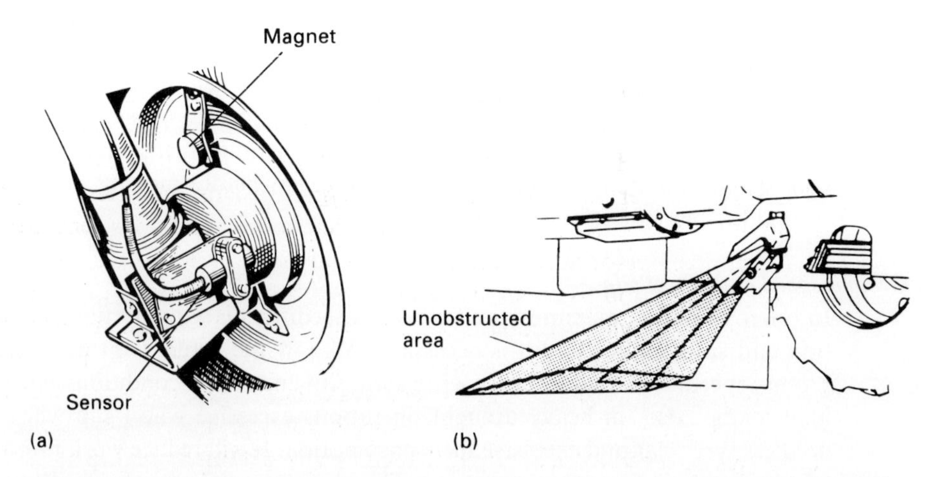

Figure 3.7 Ground speed measurement. (a) Rotation sensor on a non-driven wheel (RDS Technology). (b) Mounting position and *footprint* of a Doppler speed sensor (Dickey-john).

Wireless Telegraphy Act of 1949, an annual licence fee is imposed by the Department of Trade and Industry, which is responsible for radio regulations. The two operating frequencies in general use are 10.587 and 24.125 GHz, but 24.3 GHz in the UK.

Physically, the speed sensor is a cylindrical, wedge-shaped or rectangular enclosure with a blind face (see Fig. 3.7(b)), through which the radiation is transmitted and reflected radiation received. It is mounted on the tractor chassis, between the wheels and up to 1 m above the ground. The objective is to locate it close to the centre of gravity of the machine, where pitching movements of the latter will contribute least to the relative movement between the sensor and the ground. The unit is mounted so that its axis of radiation is at a closely specified angle to the ground, normally between 35° and 45° from the horizontal. It is usual to direct the radiation rearwards, rather than forwards, to protect the sensor against stubble and other mechanical hazards. The beam axis must be within ±2° of the specified angle and the whole *field of view* of the sensor must be unobstructed to avoid reflections from anything but the ground surface.

The *footprint* created at the soil surface depends on the beam angle, of course. The smaller this is, the better, because the meter measures the frequency change of the radiation reflected back into its front window from the ground and the change depends on its angle relative to the horizontal, as well as the ground speed of the vehicle. Beam spread makes it inevitable that reflections will enter the sensor from a range of angles other than that of the specified beam axis, therefore the Doppler frequency shift covers a band of frequencies, rather than a single value. A smaller footprint reduces this frequency scatter, which limits the accuracy of the measurement. An advantage of the 24.125 GHz meters is their smaller beam angle – in broad terms about 10°, compared with 20° for the 10.587 GHz meters. Frequency scatter also increases with surface roughness, although this is a smaller effect, and the amount of radiation reflected back to the meter is increased at the same time.

The amount of the central frequency shift (i.e., that due to the Doppler effect on the axial beam) is very small. The value for the 10.587 GHz meter is about 16 Hz/km/h (26 Hz/mph). The corresponding, pro rata figures for the 24.125 GHz meter are 36 Hz/km/h (58 Hz/mph). This indicates the greater sensitivity of the latter. Electronically, the microwave radiation is generated by a small semi-conductor device known as a Gunn diode, which only requires a low voltage energisation to do so. The output power is about 5 mW, beamed via a conical antenna which also collects the return signal. A small fraction of the transmitted signal is *mixed* with the return signal. However, the resulting frequency-difference signal is often very weak, as well as indefinite in frequency. The meter therefore requires microprocessor-based intelligence to extract a square-wave, speed-related output that can be passed on to the display unit. The microprocessor can also adjust the sensitivity of the system automatically, so that when the vehicle is at rest the sensitivity is low and the meter ignores weak signals generated by machine or crop movements. The speed of response of the meter can also be software controlled to suit field conditions. In this case the microprocessor is programmed to provide a long time constant when the machine is operating at or near constant speed, which eliminates unnecessary fluctuations in the displayed value of ground speed. Sudden changes in speed automatically switch the meter to fast response.

The accuracy of Doppler speed meters depends not only on their microwave frequency but also on speed and surface conditions. As a general indication, above 2 km/h they are accurate to ± 2 or 3%. The meters can be calibrated in the same way as the wheel sensor, using a level field course marked out as accurately as possible.

Situated as they are in the open and close to other electrical and electronic equipment, these meters need an enclosure with high environmental integrity. Anti-vibration mounts help to reduce the effects of vehicle-induced vibration and the casing is made sufficiently strong to withstand the mechanical knocks that field equipment may suffer in normal use. The enclosure is sealed to exclude water and chemicals. Electromagnetic compatibility (section 2.4) is maintained by careful screening. Protection against battery overvoltage is also provided. During installation care must be taken to fix the screened cables where they will be least vulnerable and to ensure that the meter is not mounted at a point of high vibration. The latter can be checked by observing whether there is a non-zero reading when the vehicle is stationary.

As far as the operator is concerned, the 5 mW power radiated presents no hazard but users are advised not to look at the face of the sensor directly while it is in operation, to avoid any possibility of eye damage.

Wheel slip (ws) is normally presented as a percentage ratio, in the form

$$\mathrm{ws} = \frac{D - d}{D} \times 100\%$$

where D is the distance that would have been covered in a given time in the absence of slip, and d is the actual distance covered in that time. This converts to

$$\mathrm{ws} = (1 - \frac{d}{D}) \times 100\%$$

$$(d = D, \mathrm{ws} = 0\%; d = 0, \mathrm{ws} = 100\%)$$

The range of particular interest is from 10 to 25%, in general. For traction purposes and acceptable loading of the transmission, 10 to 15% is recommended for 2 WD tractors and 2 or 3% slip less on 4 WD machines. Above 15%, tyre wear and fuel consumption increase with diminishing gains in work rate. At 25% slip, these could increase by 40% and 10% respectively, with consequential effects on maintenance costs and service intervals. Unfortunately, wheel slip is not easy to detect by eye below 25%, hence the need for an operational aid.

The output from either of the above ground speed sensors can therefore be used to measure wheel slip through simultaneous measurement of the rotational speed of a driven wheel with a known circumference ($\pi \times$ its effective diameter).

This requires a second sensor on the driven wheel, of course. The ground speed sensor measures the value of d per unit time and the driven wheel sensor provides the corresponding value of D. From their ratio a meter can easily compute and display percentage slip.

Auxiliary machine controls

Once wheel slip has been measured it can be used to limit slip. This facility was incorporated in some of the many work rate monitors for tractors that were developed in the 1980s. Figure 3.8 shows a good example of these monitors. The control knob, K, selects any of the 16 functions shown, for display on the LCD panel. [*Note* The cost factor is a notional figure derived from measured fuel consumption, time of operation and engine speed, representing fuel, labour and maintenance costs, respectively. This was introduced to help the driver to operate in an economical way.]

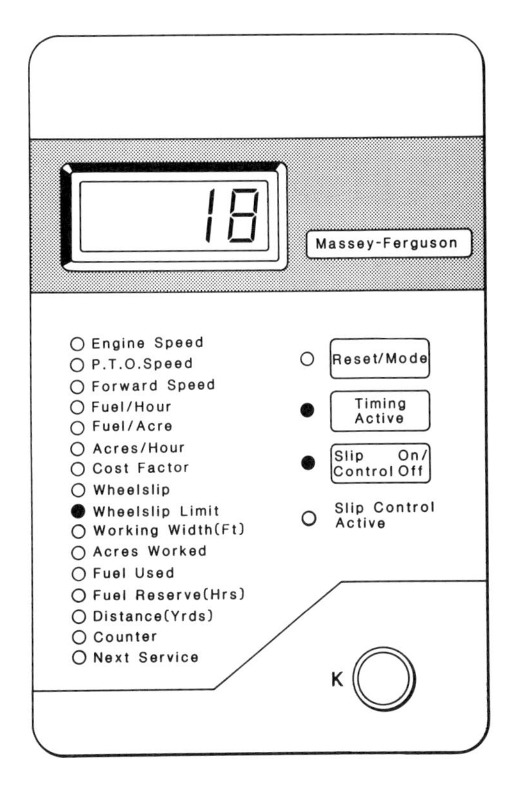

Figure 3.8 In-cab monitoring and control unit (Massey-Ferguson).

Wheelslip Limit is shown as the current selection (the associated LED is *on*) and the setting is shown as 18(%). This and other settings are programmed in with the aid of the Reset/Mode button and K. The Slip Control button allows a wheel slip control system to be engaged or disengaged. When it is engaged it temporarily overrides draught control settings (i.e. raises the implement) when the slip exceeds a preset limit. If this happens the LED associated with Slip Control Active is lit. Setting of zero slip requires a run with the tractor at normal working speed and with the implement raised. Then, pressure on the Reset/Mode button for at least 4 s establishes the zero slip reference. Two touches on the same button allow the function selector knob to adjust the wheel slip limit setting in steps of 3% (the display shows the value selected). Another touch on the button causes the monitor to memorise the new setting. The recommended way of selecting the limit is to note the wheel slip when the tractor and implement are judged to be working normally and to set the limit about 5% above that level. The automatic reaction of the wheel slip control system is very rapid, which greatly reduces the risk of the tractor becoming bogged down in difficult conditions. However, if the Slip Control Active LED is lit for much of the time the operator may decide that the wheel slip limit should be increased.

The same manufacturer simultaneously introduced a complementary system for 4 WD tractors, designed to free the operator from repetitive control tasks. The electronic logic system provides electrical control of 4 WD, differential locks and three-point linkages, semi-automatically.

Thus, if the driver engages diff-lock, via the appropriate button, the system automatically disengages it when the three-point linkage is raised and re-engages it when the linkage is lowered. This prevents oversights at headland turns in particular. Braking also disengages diff-lock, thereby facilitating rapid turns. Disengagement is automatic, too, if the tractor is travelling at more than about 14 km/h, making the tractor safe for road work. In this case, the diff-lock is not automatically engaged when the tractor's speed falls below 14 km/h. The driver can always see when the diff-lock is engaged (the push-button's indicator light is on) and the automatic control can be overridden via the button.

At road speeds over 14 km/h 4 WD is automatically disengaged, to avoid unnecessary tyre wear and fuel consumption. However, it is automatically re-engaged if the driver brakes above the same speed, to improve braking performance and control, and it will remain engaged until the brakes are released. Selection of diff-lock automatically engages 4 WD, but not vice versa, and subsequent deselection of the diff-lock does not mean that 4 WD will be disengaged. A light also indicates when the drive is engaged, except when it is augmenting the brakes, and it is possible for the driver to override the system by continuously holding down the 4 WD push button.

The system also protects the electrohydraulically operated gearbox range shift, to prevent a shift if the tractor's speed is too high, as well as ensuring that when the engine is started the shift is in the same state as it was when the engine was stopped. Further, if there is a creeper gear, it will only allow the gearbox to be in low range when creeper speeds are selected.

The system can also regulate an independent pto (IPTO), to avoid excessive acceleration or clutch slip at light or heavy loads, respectively. This is achieved by control of the hydraulic pressure in the IPTO's clutch, to produce constant acceleration, independent of load. If the implement is blocked and the driver tries to engage the pto, the absence of acceleration causes the transmission control to disengage the clutch. Excessive clutch slip produces the same result. Either potentially damaging condition causes flashing of the light that is normally *on* when the IPTO is selected. The driver must then deselect the drive, which is engaged via a lever rather than a button, to reduce the risk of accidental operation. The system will also prevent pto overspeed by monitoring engine speed and disengaging the pto if it is too high. Hydraulic pressure monitoring is under system control, too, to avoid needless operation of the associated warning light. This should be extinguished a few seconds after the engine is started. Otherwise, if the light shows for more than 2 s the control will disengage those functions at risk.

On the road, when a tractor is carrying a heavy, mounted implement the bounce that can arise at higher speeds is a cause of driver discomfort and it constitutes a safety risk. Therefore some manufacturers have marketed active transport control systems, which can be engaged when required to alleviate this problem. They apply damping to the shock loads at the rear axle, controlled electrohydraulically, to even out the bounce.

Implement monitoring and control

Here the role of instrumentation and control systems is reviewed from primary cultivation to harvesting, including self-propelled harvesters as well as tractor-drawn implements. Guidance systems are dealt with in the following subsection.

Equipment of this nature is not often found on cultivation equipment although the cultivation technique employed at any time may have been based on surveys of soil density, as described earlier, or by tests with penetrometers – some of which are equipped with load cells and linear displacement sensors, to automate force/displacement measurement. However, electronic control of electro-hydraulics for manoeuvring heavy cultivators has been introduced commercially in the 1990s.

A good example is the control system for a multifurrow, reversible, semi-mounted plough (Fig. 3.9) which is designed, *inter alia*, to remove a stressful task from the operator at headland turns, as well as reducing the space and time required for the turn. This has a control unit in the tractor cab, an electro-hydraulic control unit on the plough's headstock, coupled to the tractor's own hydraulic system, and a variety of sensors. The in-cab unit enables the operator to set the desired front furrow width, the ploughing depth and the offset. Then, in automatic mode, when the tractor reaches the headland the operator has only to operate the tractor's hydraulic lift to initiate an automatic turning sequence. The lift actuates a switch between the main plough beam and the headstock. Then the tractor has to move forward for a preset number of revolutions of the rear wheel, sensed by a proximity switch. The number depends on the length of the plough. At the end of the counting period the sequential control programme instructs the plough's electrohydraulic unit to lower the wheel, thereby lifting the plough out of work. The offset is then closed, to improve stability during the turn, and the

Figure 3.9 Semi-mounted plough with steering control (Dowdeswell).

rear wheel takes a 45° steering angle. Here rotary potentiometers are used as angular sensors at the pivot point and the rear wheel. Then the plough is reversed and another proximity switch is actuated when that operation is complete. That actuates the next step in the programme, which is restoration of the offset to the programmed level, sensed by further position sensors. The operator steers the tractor round the headland turn while this sequence takes place.

When the front wheel of the tractor is in position for the next pass the operator presses a button on the control box and the rear wheel reverts to its normal position. At the headland marker the tractor linkage is lowered, the rear wheel pulses are counted again and at the end of the preset count the plough returns to its preset working depth. The pivot headstock allows 180° rotation, which makes it possible to turn sharply at headlands. In manual mode the steered wheel can also be used for easier negotiation of corners and obstacles, as well as counter-acting crabbing during work across sloping land. It is also possible to set the front furrow to a minimum width, so that it runs empty behind the tractor's rear wheel. This reduces draught when required – say, for working up steep slopes.

An important feature of this system is the design of the hydraulic valve block on the headstock unit, which enables it to work with any of the three hydraulic systems found on modern tractors – i.e., open-centre, closed-centre and load-sensing. The possibility of a mis-match between the tractor and implement hydraulics can therefore be avoided and risks of serious malfunction eliminated. Changes between these three systems can be made by the operator, according to the tractor's requirements. In addition, the integrity of the electrical wiring is safeguarded by running it either within the plough beam or in hydraulic hoses.

Operator safety is also addressed. To initiate the system, the operator must first release a lock on the control unit, using a key. Without the key, the plough is safe to work on. In addition the automatic sequence can be overridden at any time, if required.

Another manufacturer has incorporated a height sensor for continuous monitoring of the depth of the rear wheel as a means to maintain more constant ploughing depth. The output from the sensor is compared with the upper link setting, to provide an error signal which is corrected by the hydraulic lifting ram at the wheel.

Precision drilling and planting call for constant ground speed or adjustment of sowing or planting rate in accordance with that speed. Since precise control of ground speed is not generally possible in the field multirow seed drills are now available with an electronically controlled drive to the metering shaft, for regulation in this way. However, a more common device on these drills alerts the operator to any drill blockages. The usual sensor is a small photodetector illuminated by a small light source (e.g., LED) in each delivery tube, which during drilling produces a succession of pulses as the light is interrupted by the passage of seed. A corresponding array of indicator lights on the in-cab monitor provides warning of a blockage in any of the channels. This is usually supplemented by an audible alarm. When some channels are deliberately closed for tramlining (see later in this section) the alarm system must be capable of ignoring them, of course.

As with any photoelectric system, light sources and detectors need to be free of heavy contamination if they are to operate reliably. Some chemical seed treatments have been found to affect the above system. The prescribed remedy is to decontaminate them with a kitchen cleaner, followed by a water rinse.

An in-cab monitor that allows a *population count* of the seeds planted per row and per ha is shown in Fig. 3.10. The user sets up the row spacing required and the intended planting speed, before drilling. The information is entered into the monitor's computer via push buttons and registered by a digital panel. Then, during operation, the same display continuously shows the average seeds/ha

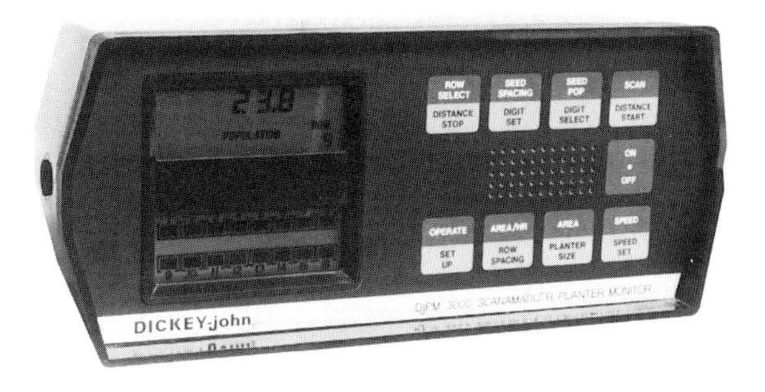

Figure 3.10 Planter monitor (Dickey-john).

being planted by all the rows. Further push buttons allow the operator to monitor any selected row, or to scan all rows, in order to compare their planting rates. Visible and audible alarms warn the driver of malfunctions, which can include a seed blockage or an empty hopper. When the monitor is coupled to a ground speed sensor, it provides actual planting speed. Area planted and plant rate can also be displayed in this case.

Planting in accordance with ground speed has found application in automatic potato planting, in various forms, since the 1980s. An early commercial form of this system employed microprocessor control to planting at speeds up to 12 km/h. It could plant almost 12 potatoes/s/row, with low damage to chitted seed, using planting belts driven by a variable-speed motor. A wheel revolution sensor on a land wheel provided a ground speed signal to the microprocessor, which enabled it to control the speed of the belts to match that speed. In that way the tubers were planted at equal spacing. Push buttons on the control unit allowed the operator to select tuber spacing (in the row) between 100 and 500 mm, in very fine (10 mm) increments. A more recent machine also incorporates facilities for the operator to prime and unload the belts and to ensure that seeds at row ends are fully covered with soil by the machine's covering discs or ridgers.

Similarly, ground speed measurement has been applied to control of the distribution of granular fertilisers by twin disc spreaders. The actual flow rate can be measured by suspending the hopper on one or more load cells. An ECU can then adjust the rate to the required value by reference to the disc and ground speeds. Nevertheless, the uniformity of application also depends critically on the choice of bout spacing to achieve the optimum degree of overlap of the fringe distribution on adjacent bouts. The overall accuracy therefore can only be determined by calibration (static or otherwise) of the profile of the lateral distribution, obtained with the type of fertiliser to be spread.

However, boom sprayers were the first field machines to be fitted with commercial application rate controllers based on ground speed measurement. These controllers were marketed in the early 1980s as part of the response to accumulating evidence that the variability of spray application was unacceptably high. In fact, a UK survey in the late 1960s had revealed point-to-point variations in some circumstances of up to 40:1 ratio. Only part of that variability was due to variations in ground speed, of course. Boom sway and yaw, spray drift and users' failures to check – and, if necessary, replace – worn nozzles, were potent sources of inaccuracy. Since then, boom suspensions and other improvements of boom design have done much to reduce the effect of sway. Equally, a variety of new spray nozzles and spinning discs provide closer control both of the discharge pattern and the droplet spectrum, to increase the effectiveness of the spray and reduce spray drift. Thirdly, the establishment of tramlines matched to boom widths at seed drilling has greatly reduced overlapping and underlapping of spray deposits on adjacent bouts.

Despite these improvements, every means to improve the precision of spray application is important, both financially and environmentally. Even if patch spraying (see section 3.4) becomes a generally used technique, ground speed controlled systems such as that shown in Fig. 3.11(a) may well continue to provide the best application method available to the spray operator. It will be seen from the figure that the control unit can accept a ground speed signal from either a land wheel or a Doppler sensor. A pressure or flow sensor provides the required information on the application rate and the control unit adjusts the rate to the amount determined by the ground speed, via the pressure control valve. The target application rate is set by the operator, using the up/down keys on the control unit.

The control unit also acts as a data processor of the form widely known as a work rate meter. Its analogue and digital displays show ground speed, spray pressure, application rate, volume applied, total and part areas treated, remaining tank contents, target speed and pressure operating limits. Data can be downloaded to a printer or PC, to provide information necessary for compliance with safety regulations, *inter alia*. The power supply and all inputs/outputs are protected against transients to an SAE standard (J1211). Its temperature range is –20°C to +50°C. The pressure sensor unit's application rate is not sensitive to the number of boom sections in use (see below), and it can be preprogrammed for up to four different nozzles.

There are several methods of measuring the liquid flow rate but most systems employ either a pressure sensor or a turbine flow meter. Sensors of the former type measure the discharge rate via the pressure in the spray line, which typically lies in the 100 to 400 kPa range. The non-linear relationship between the pressure and the discharge rate (the rate is approximately proportional to the square root of the pressure) is accommodated by the controller's computer. With pressure sensing the discharge rate at each nozzle is independent of the number of nozzles in action, so one section of the boom can be switched off without significantly affecting the remainder. The same applies if a nozzle becomes blocked during the operation. The required range of the flow meter depends on the boom width and the maximum rate of application that it must accommodate. For a 10 mm boom, delivering a high volume of 1000 l/ha, at a standard ground speed of 6.4 km/h, the meter must cope with approximately 110 l/min.

A feature of hydraulic nozzles is that a pressure range of less than 3:1 is all they can tolerate before their spray distribution pattern and drop size spectra are altered significantly. This corresponds to a change in discharge rate of about 1.6:1, therefore there is a limit to the range of ground speeds that can be accommodated by a fixed set of nozzles if these unwanted changes are to be avoided. Most controllers allow the user to program in preset operating limits. Then if the sensor output indicates an out-of-limits discharge rate the driver is automatically warned, audibly and visibly.

However, the above system may not cope in all circumstances with the effect of pressure regulation on the drop size spectrum created, especially at higher ground speeds. This has led to the system in which water is discharged at a steady rate and the chemical is mixed with it in amounts dependent on the ground speed. Figure

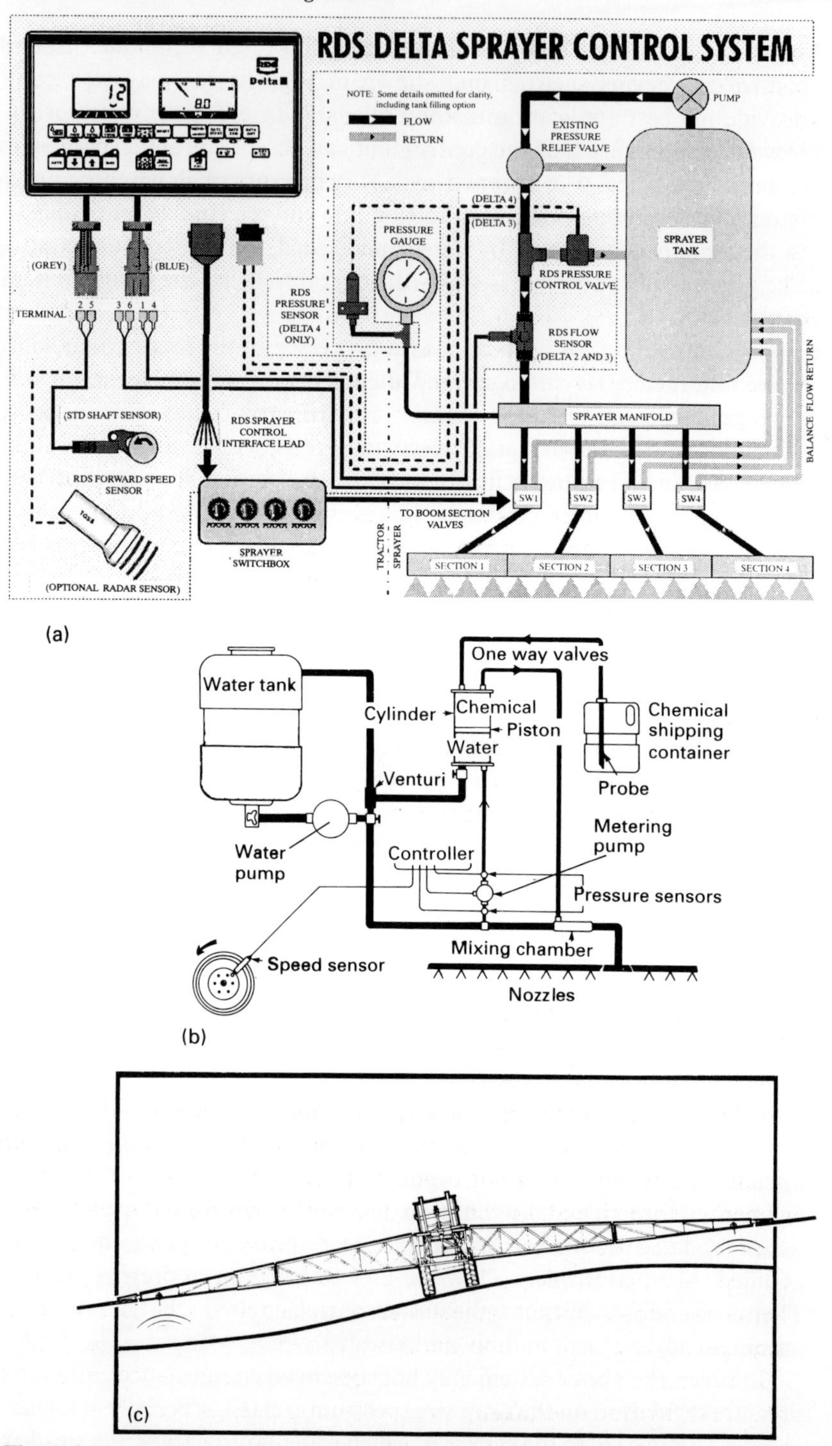

Figure 3.11 (a) Spray-rate control (RDS Technology); (b) spray concentration control (Silsoe Research Institute); (c) active boom suspension (J.W. Chafer).

3.11(b) shows a form of variable concentration control. The cylinder with its close fitting piston is first filled with water via the metering pump line, then steadily drained by Venturi action as water from the main tank is circulated by the water pump. This lowers the piston and draws the concentrated chemical into the upper part of the cylinder. The valve leading to the Venturi tube is then closed. When spraying begins the controller reads the input from the wheel speed sensor and controls the flow of water which is taken from the pipe leading to the mixing chamber and fed to the cylinder via the metering pump. In consequence, the piston again rises and an equal amount of neat chemical is driven into the mixing chamber before discharge through the nozzles. An additional and valuable benefit is that there is no surplus of diluted spray to be disposed of at the end of spraying. Any concentrated chemical remaining in the cylinder can be recycled to the shipping container.

As stated earlier, hydraulic nozzles suffer from inevitable wear which can affect their discharge characteristics, thereby undoing much of the advantage attainable with precision sprayers if undetected and uncorrected. However, simple, hand-held instruments are available for rapid checks of these nozzles, individually. They incorporate a turbine flow meter below a collector which fits over the nozzles, and when the appropriate button is pressed the flow rate is displayed digitally in either l/min or gal/min at rates up to 4.5 l/min or 1 gal/min, respectively. The accuracy is about $\pm 2\%$ of fro (full range output). With the aid of this tool faulty nozzles can be identified and replaced before they markedly affect the uniformity of spray deposition.

Another diagnostic aid is provided by the spray controller with programmed nozzle calibrations in its memory. This can draw attention to abnormal discharge characteristics which point to poor spray quality, meriting inspection of the equipment. Some air-assisted sprayers employ a memory facility in another way. The controller can be programmed with preset combinations of air and liquid pressure, to achieve spray droplet spectra from very coarse to fine, as well as memorising the characteristics of standard and liquid fertiliser nozzles. Colour-coded touch buttons on the controller's keypad match the colour coding of the nozzles, to simplify system set up.

Tracking systems for trailed sprayers are another development of the 1990s. They were introduced to take over from the operator the task of turning with them at headlands. When the tractor begins to turn at the headland the angular misalignment between it and the sprayer is detected by sensors at the tractor end of the drawbar that connects them. The sensors can be a rotary potentiometer or a pair of displacement sensors, one at each end of the drawbar. These actuate an electrohydraulic ram, which steers the sprayer so that it follows the tractor wheels. This system can also be used to stop the sprayer from crabbing on side slopes.

In connection with side slopes, reference was made earlier to the advances in boom control that have taken place. Although these have been due mainly to passive springing and damping, active boom suspensions have been marketed for wide boom sprayers. Ultrasonic, downward looking, ranging devices have been mounted near each boom tip to measure the height above ground level at both extremities. The difference between the two measurements provides an error signal which is fed into a closed loop control system. The error is minimised via an electrical drive which adjusts a suspension link. The system is particularly useful for work on side slopes, where a positive drive is necessary to maintain the boom parallel to the slope. This point is illustrated by Fig. 3.11(c).

The transducers are of the widely used pulse-echo type, which must be mounted in a waterproof housing and protected as far as possible from spray deposits. Short bursts of energy at a low ultrasonic frequency are transmitted at brief intervals (60 ms in the case of the system shown in Fig. 3.11(c)). Between these bursts the transmitter/receiver has to collect the energy reflected from the ground, so that the time difference can be calculated and converted to distance. The time difference is short, because the velocity of sound in air is over 330 m/s at temperatures above 0°C, so the echo will return in about 10 ms in most circumstances. It is also not clearly defined, because reflected energy is received from a fairly wide area. Therefore extraction of the distance between the transmitter/receiver and the nearest point on the ground requires some signal processing. Fortunately, the fact that the velocity of sound in air is not constant with temperature (it increases with temperature at about 0.6 m/s/°C, see Table 3.2) is not important, since the measurement is a differential one. In fact, an error signal of ±100 mm is quite acceptable in this application. On the other hand, when ultrasonic gauges are used to measure the contents of spray tanks the need for temperature compensation is evident from Table 3.2.

Table 3.2 Variation in the speed of sound with temperature.

Temperature (°C)	Dry air	Pure water	$3\frac{1}{2}$% saline water
0	331	1402	1450
10	337	1447	1490
20	343	1482	1522
30	349	1509	1546

Moving to the end of the crop production cycle, a considerable amount of harvesting work is done by the cereals combine and other self-propelled harvesters. They are dealt with later in this section. Here, as far as tractor-drawn harvesters are concerned, electronic instrumentation and electrohydraulic control have been introduced to assist the operator of big balers for straw or forage. From the tractor cab the operator can control the wrapping of these bales – setting their size, adjusting twine spacing, or switching from twine to net wrapping. The programmable in-cab display/control unit provides information on the development of the bales, including bale pressure and the feed rate into the bale chamber; diagnostics and bale counting. Balers (and other machines) can also be fitted with a torque sensor in the driveline shaft, coupled with an instrument in the cab that gives visual and audible warning when the torque is close to overload conditions. The sensor is in the form of a pair of strain gauges, mounted on the outside of the shaft with their main axes at 90° (see section 1.5). When strained their output controls the current drawn from a 12 V supply by an amplifier unit inside the shaft, to which they are connected. The 12 V supply is fed to this assembly via standard slip rings with carbon brushes, The externally measured current (in the mA range) provides an indication of torque in the driveline. Some silage wrappers also have electronic control of the film feed and tensioning.

Pick up of metal objects by forage harvesters generates a double risk – first to the machine and second to an animal which might ingest it. The latter risk is particularly associated with the pick up of small pieces of wire or nails, which is not an uncommon occurrence and not easy to monitor. Nevertheless, sensitive detectors of ferrous metal objects were developed by Sperry–New Holland in the late 1970s, and introduced as front-end additions to the input feed roll of their harvesters in the 1980s. The original form of the system is shown in Fig. 3.12. The sensor is an electrically balanced assembly of two permanent magnets in a non-magnetic (aluminium) frame, and a coil in a figure-of-eight shape. It is mounted in the lower of the two feed rolls, both of which are non-magnetic. With this configuration the crop material is constrained to pass close to the sensor, in compacted form, as it is taken in by the rolls. Hence, the presence of a ferrous object in the material at that point has a significant effect on the magnetic field created by the sensor. Any asymmetry in the effect on the two halves of the sensor as the object passes through the rolls immediately unbalances the system electrically and the coil produces an output signal proportional to the object's size and speed.

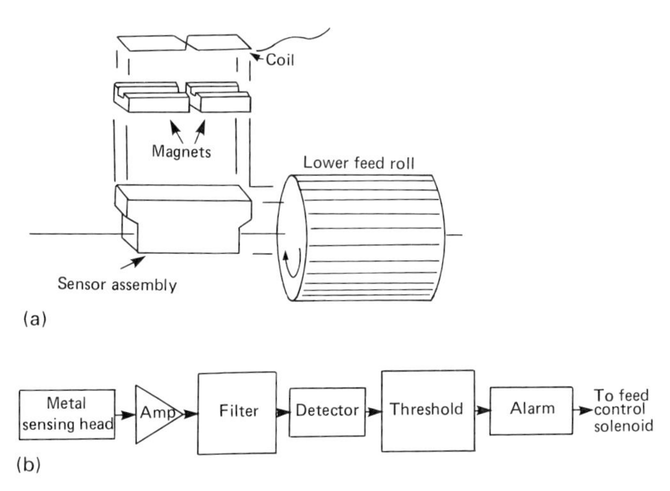

Figure 3.12 Tramp metal detector for forage harvesters: (a) sensor assembly; (b) signal processing (New Holland).

The low-level signal first needs amplification before it passes through a special filter stage. This suppresses noise generated by the machine, enhances the tramp metal signal and removes its speed dependence. The signal at this point still fluctuates in amplitude between positive and negative, in relation to zero voltage. The next stage, labelled 'Detector' in the diagram, removes one half of the fluctuations, to create a *unipolar* signal, which can be compared with the voltage threshold level in the following stage. In the absence of tramp metal the signal (which is due to residual noise) does not exceed the threshold level, but when metal is picked up the threshold is exceeded and the alarm circuit is activated. This actuates a ratchet and pawl mechanism, which shuts off the feed roll drive and puts the main gearbox into neutral. The time taken for this response is about 50 ms, which prevents the ferrous object passing through the rolls. The alarm circuit also alerts the driver to the need for action – which entails reversing the drive to expose the tramp metal, shutting down the harvester and removing the cause of the stoppage. A fail-safe circuit ensures that the harvester cannot be operated if the sensor is disconnected. It also alerts the operator in the event of any malfunction of the monitor.

In its later form the detector employs two stainless steel front rolls. The magnets and their associated electronics are located between the two rolls. The magnets have been repositioned on offset planes to broaden the magnetic field. Overall the later design provides more uniform and greater sensitivity across the feed roll throat area, much less dependence on the orientation of the ferrous object, and earlier detection of its presence. The reliability of the system has been further increased by placing all of the electronics in the lower roll. This avoids the transmission of weak signals to an external amplification and detection unit. Aluminium shielding protects the electronic assembly from EMI, which can cause false alarms, and the lower roll is completely closed, to protect the assembly from moisture and corrosive plant juices.

Integrated systems

Step by step the tractor and implement industry is moving towards the integrated monitoring and control system shown in Fig. 3.13. In effect this is a SCADA system (Fig. 2.5) with the operator and the central processing unit (CPU) acting as clients and the remainder of the system comprising the servers – all interacting with each other through the data transfer network. The microprocessor-controlled engine, transmission and implement control systems already described in this section will then be placed under the general supervision of the CPU, which may modify their control algorithms or employ their output to modify the algorithm of another subunit. Overall, the CPU can provide information about the state of the system, or any subsystem, and it can advise the operator of the choice of settings. In reverse, the operator can programme and reprogramme the system via the CPU. Both can interact with the external client – the farm office management system – via the data input/output device, which is currently likely to be an environmentally protected printer or a hardware data transfer device

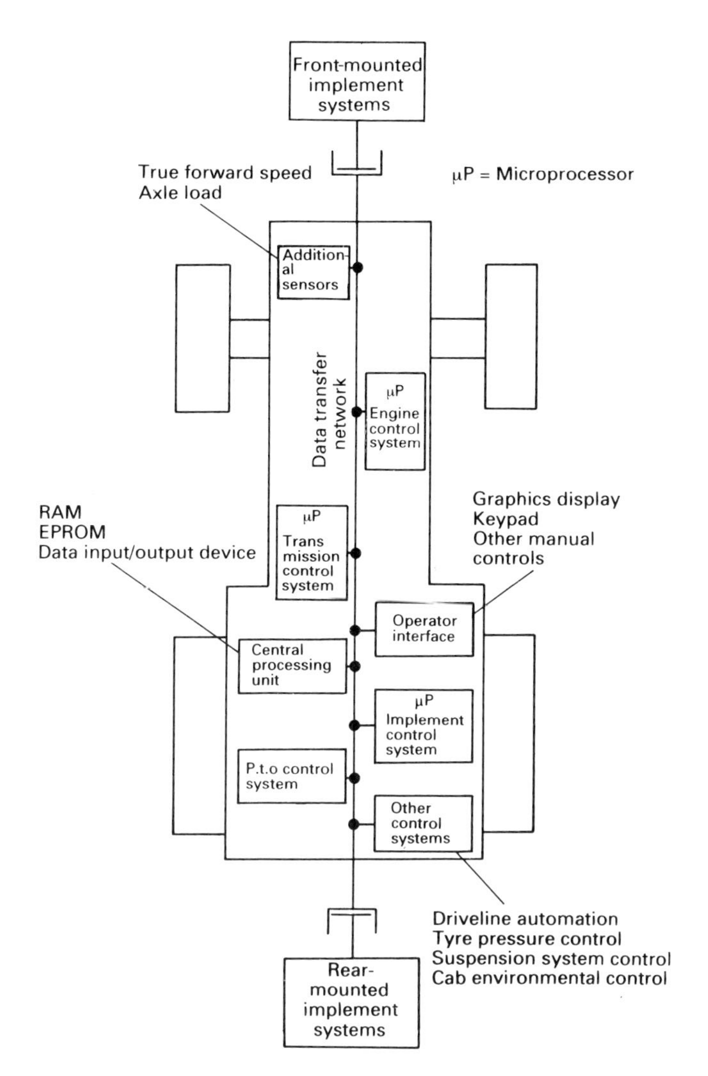

Figure 3.13 Proposed integrated control system for an agricultural tractor (Silsoe Research Institute).

such as the PCMCIA card reader/writer. These do not provide real time communication of information and instruction between the machine and the central system, of course: that is usually reserved for voice communication by radio telephone. However, remote supervisory control of robot vehicles is already a reality in some areas of industry and these techniques may find application in crop production, given sufficiently economic and reliable methods of position fixing or navigation by machine vision on field machines.

Position fixing and navigation are discussed further in section 3.4 and an example of the development of machine vision, coupled with artificial intelligence, is given in Chapter 4. Here it is sufficient to note that the front-mounted implement of Fig. 3.13 might be a forward-looking vision system for tractor guidance, and that the input/output device might deliver real-time positional data for site-specific soil or crop treatments.

Undoubtedly, one important function of the CPU will be to employ engine performance maps to look for optimal operational modes, such as maximal economy or maximal performance. These multidimensional maps relate engine speed and torque, fuel consumption and work output in ha/h. Finding optima in such complex space is a task for the search algorithms referred to in section 2.3. However, simpler versions of this facility have been available on commercial tractors since the mid 1980s. Figure 3.14 is an example of the relatively simple mapping employed for calculation of the gear for greatest fuel economy in particular field conditions. To obtain this advice the operator presses the ECO (economy) button on the tractor's monitoring and control unit. This causes the CPU to register the engine and wheel speeds, from which information it deduces the gear that is engaged. The output of an exhaust gas temperature sensor is registered at the same time. The system then works on the basis that the gas temperature is closely related to the engine torque, so allowing the torque to be derived from the temperature. Taking the example in Fig. 3.14, this establishes that the engine is operating at point 1 on its power (torque × speed) curve. The CPU then refers to look-up tables in the computer's memory, seeking a point of equal power but lower specific fuel consumption. In this example it will find it at point 2. Then it signals to the operator that it has completed its search. Another press on the ECO button initiates display of the new engine speed and recommended gear ratio.

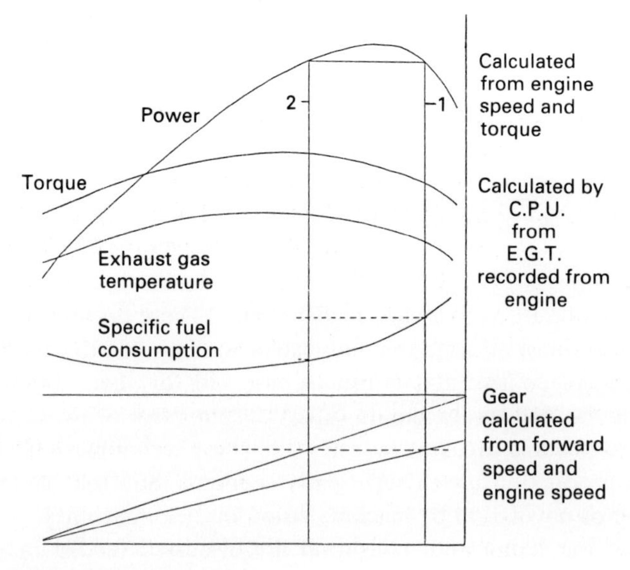

Figure 3.14 ECO system (Renault).

Diagnostic monitoring of engine performance by reference to sensors for measurement of temperature, pressure, flow, speed, vibration and emissions is also likely to become a standard feature, thereby simplifying – and encouraging – routine service checks.

Much of the foregoing is a prediction of future developments but the central feature of the integrated system in Fig. 3.13 – the data transfer network – is already taking shape, as stated in section 2.2, under controller area networks.

The CAN data transfer protocol has many facets and its application to agricultural/implement combinations raises many issues that will not be settled quickly – not least harmonisation of the German DIN and the American SAE approaches with the evolving ISO standard. However, a general outline can be given here.

The CAN is based on a number of stations, connected to nodes on the bus, as in Fig. 3.13. All stations are equal – i.e., there is no central master – and each can communicate with all of the others, as with the Ethernet (section 2.2). They do not have addresses. Their outgoing messages contain an identifier which indicates the nature of their contents, coupled with a priority ranking which is involved when there is competition between stations for use of the bus. Priority arbitration is crucial because CAN is working with real-time control, requiring data transfer rates up to 1 Mbit/s, like Fieldbus (section 2.2). Therefore an urgent message from, say, a wheel slip monitor must take precedence over a more slowly varying parameter such as engine temperature.

The structure of a message in standard format is shown in Fig. 3.15(a). A logic 0 start of frame is followed by an 11 bit identifier and a one bit remote transmission request (RTR). These constitute the arbitration field. The RTR bit identifies whether the frame is an outgoing data frame or a request frame. RTR = logic 1 indicates a request. The ensuing control field starts with an identifier extension bit, which indicates whether the identifier contains more than the 11 bits of the standard format. That is followed by a reserve bit (r_0) and a four-bit number indicating the number of data bytes in the data field. The penultimate field is for the cyclic reducing check (CRC) mentioned in section 2.2 as a means for error checking. This is particularly important in the vehicle context because of the likelihood of corruption of data by EMI. CRC operates by adding redundant check bits at the transmitter end. At the receiver the check bits are recalculated – using the same procedure as that employed by the transmitter – on the basis of the information bits received. If these do not tally an error is diagnosed and retransmission is requested. This field also includes a frame check which looks for format errors. Last, the two-bit ACK (acknowledge) check detects transmission errors revealed by the failure of the transmitter to receive an acknowledgement from its targeted recipients. The error checks are augmented by bus monitoring by each node during transmission, to look for disturbances. If these persist CAN has the ability to pinpoint statistically any station that appears to be the cause of errors. This can lead to self-deactivation of that station.

The arbitration between stations seeking access to the bus is done through a

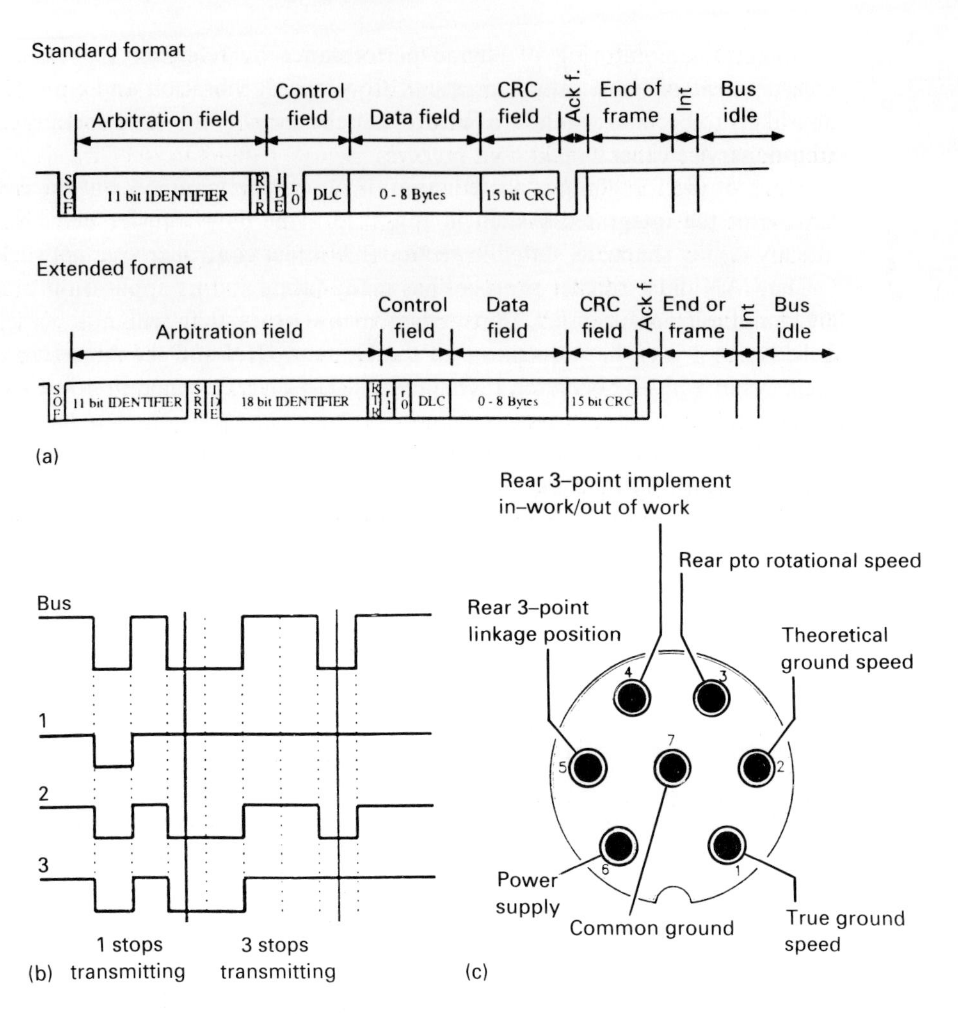

Figure 3.15 CAN system: (a) standard and extended formats; (b) arbitration; (c) ISO 11786 connector [*Note* pins 5 and 6 are free on the corresponding DIN 9684 connector] (KTBL, Darmstadt).

bit-by-bit comparison of their identifiers, by a form of knockout competition. The rule is that logic 0 beats logic 1. Therefore, referring to Fig. 3.15(b), where three stations are competing, station 1 goes out at bit 3 and station 3 goes out at bit 7. In that contest station 2 obtains access to the bus. Since no two identifiers are alike there is always a decision. The losers have to try again. The consequence of this rule is that identifiers of lower value always take precedence over those of higher value. Therefore highest priority goes to the identifier starting with the digits 000.

The German LBS system (Landwirtschaftliches BUS-System) – which is part of the DIN9684 standard – defines an identifier structure designed for the agricultural application of CAN. This allocates identifiers to eight groups, ranked according to priority and identified by the first three digits of the 11-bit identifier field. The remaining eight digits in each group are allotted to specific subgroups. A further breakdown involves identifier digits in the data field. An example is given in Table 3.3. Bit A distinguishes between system logon and system management. The next four digits provide 16 classes of machine or implement. For example, 0001 identifies tractor and 0010 a primary cultivator. The last three digits (POS) provide eight classes of position. In this classification 001 identifies front mounted. Further breakdown brings in process classification, such as set-points and sampling intervals, maximum and minimum values, integral values and calibrations. The lower priority groups start with 001 (basic messages), 010 (targeted messages) and so on, as specified by DIN9684.

Table 3.3 CAN identifier group (LBS).

0	0	0	A	G	E	T	Y	P	O	S	**Initialization:** A = announce/system management, GETY = implement type, POS = implement position
0	0	1	B	A	B	O	S	E	N	D	**Basic-messages:** BABO = part identifier for specific basic message, SEND = transmitter address
0	1	0	E	M	P	F	S	E	N	D	**Targeted-messages:** EMPF = receiver address, SEND = transmitter address
0	1	1	E	M	P	F	D	I	E	N	**LBS-Services:** Direction (services > > device), DIEN = individual service, EMPF = receiver address
1	0	0	D	I	E	N	S	E	N	D	**LBS-Services:** Direction (services < < device), DIEN = individual service, SEND = transmitter address
1	0	1	x	x	x	x	T	E	I	L	**Partner systems:** TEIL = device address of the complete LBS Job computer in the partner system, xxxx = free
1	1	0	x	x	x	x	x	x	x	x	**Free**
1	1	1	x	x	x	x	x	x	x	x	**Free**

Inevitably, the agricultural application of CAN brings in many more subdivisions than are normally required in the automobile context. Therefore use of the extended identifier message may be required. This introduces the modifications to the arbitration and control fields shown in Fig. 3.15(a). The changes are that a substitute remote request (SRR) follows the 11-bit identifier, the IDE bit switches to announce a following identifier extension, and two reserve bits appear in the control field. The SRR is set at logic 1, which ensures that in a contest to access the bus it will always lose to a standard format message frame.

Equally inevitably, given so many possibilities to break down the identifier by group, subgroup and monitoring and control functions, multivendor agreement is necessary on the codes to be used. The aim is to achieve *plug and play* capability for any subsystem connected to the bus.

As with all network standards, the physical layer is the base level. Communication is in two-wire serial form with logic levels 0 and 1 represented by approximately 1 and 4 V respectively. The initial connector chosen was consciously a short-term measure, pending international agreement. It was the 7-pin DIN9684 connector, on which ISO/DIS 11786 is based, although the specified uses of the two sets of pins differ slightly (see Fig. 3.15(c)). This was seen as complementary to the DIN9680, three-pin electrical power connector which had become well established on tractors – and which had also been employed for data transfer to and from work-rate monitors and control systems. However, the DIN9684 has not been adopted globally. Massey-Ferguson introduced the company's own 16-pin connector for a variety of analogue inputs. This has three pins reserved for CAN – two lines and a separate CAN earth.

Above the base level, low-cost serial link modules are available for direct connection of sensors and actuators to the bus. These are hard wired and require no software. There is also a range of microcontrollers with an integrated CAN interface, although some only support the standard message format. The low cost of these devices is due to their applications in the automobile industry, of course.

At the man-machine interface, Germany and the USA appear to be in agreement about the form of the in-cab unit. This is, in effect, a *virtual terminal* (see virtual instruments in section 2.1) with stored windows specific to different field tasks, involving specific implements. The operator can summon them to the display panel, as required. At the lowest level these can be solely alphanumeric in form but graphic displays are available and these are to be preferred ergonomically. In Germany the interface is standardised as DIN9684, part 4. Part 5 of the same standard deals with the equally important interface between the tractor and the farm office computer. This invokes the Agricultural Data Interchange Syntax (ADIS) (ISO/DIS 11787) and it relates to levels 6 and 7 of the ISO's Open Systems Interconnection mentioned in section 2.2.

The CAN has been supported by tractor and implement manufacturers as the most economical means for improving the efficiency and precision of field operations. The ability of several modules to share information from individual sensors is regarded as a particularly useful feature of the system. Among the earliest reports from the manufacturers, Fendt have published details of their ability to meet EMC requirements with their CAN system. The bus earth (ground) was connected directly to the negative terminal of the tractor battery. All power current was routed back to that terminal via a single lead from the common chassis ground. All input and output ports were guarded against common mode interference (section 2.1) and the bus lines were fitted with a termination network at both ends (see RS422 and 485, section 2.2). However, the lines were neither twisted nor screened. The system was thoroughly tested for EMC compatibility in a specialist laboratory.

Self-propelled harvesters

Two types of these machines must serve to illustrate the range of functions that have been put under the supervision and control of electronic systems.

In the early 1990s electronic supervision and control was incorporated in a six-row sugarbeet harvester (Armer Moreau). This machine employed independent hydraulic circuits for the main transmission, topper, lifter and loader. It also had five individually powered cleaning turbines, reversible to clear any blockages. The toppers and lifters were suspended on hydraulic arms, set to selected pressures in order to achieve the required working depth. Clearly, coordinated and automatic monitoring and control of this assembly was needed, leaving the operator to steer and check the information provided by the in-cab display unit. In work, the computer continuously monitored the pressures in the topper and lifter suspensions and corrected for any deviations from the preset values, in order to maintain the intended working depth. The turbine speeds were set via the computer and individually monitored and displayed. Ground speed and area worked were also displayed. Fault diagnosis software generated visible and audible warnings of changes or problems. It also provided reminders when the next service was nearly due.

The mental and physical stresses on the operator of the grain combine harvester have long been recognised, particularly when the machine is operating over uneven terrain, when changeable weather conditions make the task urgent during favourable periods and when the crop itself has been affected by wind and rain. In addition, the operator is in no position to observe directly what is happening to the crop once it has been cut and transferred to the threshing mechanism. Therefore it is not surprising that the combine has in many ways been in advance of the tractor in adopting operator aids based on monitoring of key functions, if not in the introduction of automatic control.

Apart from engine monitors, combines are fitted with shaft speed or torque sensors, to warn of malfunctions and overloading of the machine's separation and cleaning mechanisms. However, these do not provide the sensitive indication of overload obtained by monitoring the rate of grain loss at the rear of the machine. Grain loss monitors mounted at the end of the straw walkers and sieves were developed to provide the operator with that information. Given a measure of these losses the operator can judge when to reduce speed, in order to prevent the steep increase in loss that occurs once the machine is overloaded.

The loss-rate sensor is essentially a simple device, i.e. a plate on which the seeds (*inter alia*) fall and which responds to their impact in a manner similar to that of a microphone. There the similarity and simplicity end. The associated electronics has to single out the seed impacts from the rest and that is not necessarily straightforward. Traditional filtering of the sensor's output has limitations, especially when the combine is employed to harvest a range of combinable crops with seeds of different sizes. Then the requirement is for some capability in the monitor to recognise the impact signatures of the different seeds, otherwise manual adjustments of the monitor's settings are needed whenever a different crop is being harvested, rather than simple selection by push buttons or switch. This presents an application for the fuzzy logic technique outlined in section 2.3. In fact, commercial loss monitors have normally provided fuzzy output displays, by relying on the operator to set a level of acceptable loss rate, then displaying the instantaneous rate as a relative increase or decrease in broad terms. The operator's choice of a base level can also be based on a fuzzy rule, such as IF rain is expected in the next x hours THEN increase the usual base rate setting by y.

Measurement of grain yield at the harvester itself has always been of interest as a direct means to determine the productivity of individual fields, and several forms of yield meter have been devised. These divide into those that measure the input from the clean grain elevator to the tank and those that measure the output through the discharge auger when the grain is transferred to a trailer. Both types can be retrofitted to combines, although those of the former type are now standard features on some commercial machines.

Initially, these meters were more commonly fitted to the discharge auger, either by the addition of a magnetic sensor that provided a count of the revolutions of the auger during the discharge of grain into the trailer, or by fitting a sloping, strain-gauged plate to the auger's outlet, so that grain flowed over it on its way to the trailer (Fig. 3.16(a)). The former is a volumetric method, requiring a separate measurement of the bulk density of the grain (assumed constant throughout the

(a)

Figure 3.16 Combine harvester yield and moisture measurement: (a) discharge meter (Griffith Elder).

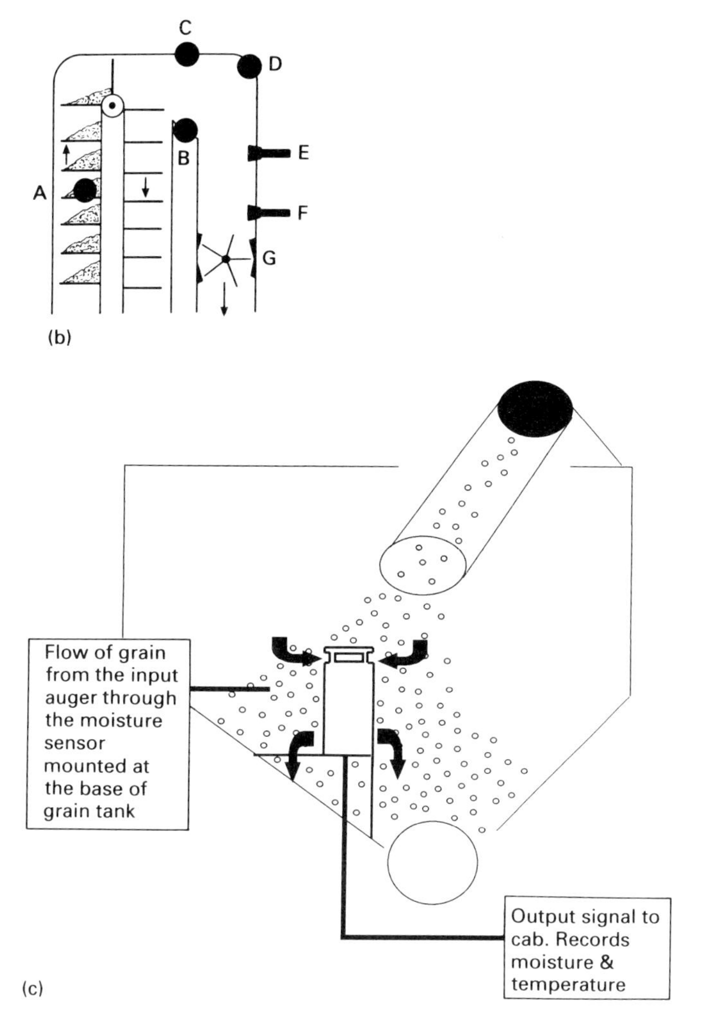

Figure 3.16 (cont.) (b) sensors on clean-grain elevator (RDS Technology; Massey-Ferguson; Claydon; Claas); (c) moisture meter (Sinar).

measurement) to convert the measurement to mass. A simple hectolitre weight balance (scales) is used for this purpose. The latter can be related directly to mass flow.

The internal meters can be divided into those that measure the volume of grain on its way to the clean grain elevator, those that measure its mass flow as the grain crosses to the down chute and those that measure its volume on the way down to the tank filling auger. Figure 3.16(b) provides a composite diagram of the three methods. On the up-travel the RDS system (A) employs an interrupted light beam to measure the height of the grain pile on each paddle. At the top the mass flow of grain is monitored continuously by one of two methods. In the

Massey-Ferguson system a low-energy gamma radiation source at B (americium 241) emits gamma rays towards a detector at C, which produces a count-rate output proportional to the attenuated radiation that it receives (see section 1.8 for radiation sensors). The only calibration required with this sensor is a measurement of its output when the elevator is running empty, i.e. the radiation received is at full strength. Since the source has a long half-life this calibration varies very slowly. In a Claas system B and D are the plates of an electrical capacitor, which forms part of an a.c., capacitance measuring circuit. Since this measurement is very moisture sensitive a separate fringe-field moisture sensor (see section 3.2, Soil and soil water) is placed in the bulk of grain below, where the bulk density of the grain is relatively constant. This enables a calculation to be made of the dry mass harvested each time the grain tank is emptied. A similar device on the tank fill auger is used with the RDS sensor, for the same purpose. However, these moisture measurements are dependent on the grain type and its temperature (see Moisture meters in section 5.3), so unless allowance is made for these factors the correction for moisture content is unlikely to be as precise as that attainable with a calibrated moisture meter. Finally, the Claydon meter employs two capacitative level sensors in the side wall of the down chute, at E and F, above a power-driven cell wheel, G. E and F control the head of grain above the wheel by starting it when the grain level rises to E and stopping it when it falls to F. In that way the volume of grain discharged by the wheel per revolution is closely controlled. The actual volume is determined by calibration. Again, it requires a measurement of bulk density to convert the readings to mass flow, and reference to the moisture content of the grain to provide a measure of dry matter or, more conveniently, *dry weight* (strictly, dry mass), at the defined storage moisture level of 14 or 15% mcwb (moisture content, wet basis).

All of these sensors provide inputs to in-cab monitors, which also take in ground speed and cutting width to provide running and total yields in t/ha and work rate in ha/h (or their Imperial equivalents) after the operator has programmed in the grain's hectolitre weight. Similarly the operator can programme in the grain's moisture content, to provide the basis for automatic calculation of dry mass. Table lifting operates a switch to cut out area measurement at headland turns or at any other time when the machine is not cutting. Correction of the area computed can also be made when the machine is cutting crop at less than full width. Strip printers and smart cards are options for the transfer of data from the machine to the farm office.

A finned moisture content/temperature measuring probe described in section 5.3 has been adapted to measurement of the moisture content of grain entering the tank of a combine harvester, thereby complementing the above measurement methods. In this application the sensor is housed in a 100 mm diameter flow regulator, designed to minimise changes in the bulk density of the grain. The assembly, shown in Fig. 3.16(c), is fixed vertically to the side of the tank by bolts or strong magnets. Grain then enters the regulator via side apertures near its top and flows out from the base. The recommended position for the sensor is as low down in the tank as possible, in the region of initial grain spill from the tank filling auger, and close to the discharge auger. The last mentioned condition ensures that grain leaves the flow regulator when the tank is emptied. Thus the system provides the combine operator with a measurement of the initial moisture content of each new tankful of grain. This information alone can determine the scheduling of combining and subsequent grain drying. The unit has been designed to meet ISO and OIML recommendations for cereals and oilseeds and has calibration for six crops (including peas and beans).

Automatic control systems on combine harvesters have been introduced for a number of reasons. For work across sloping land the cutter table's angle can be adjusted to uniform ground clearance – and hence cutting height – like the spray boom in Fig. 3.11(c). At the same time a self-levelling system to keep the body of the machine level maintains as even a flow of material as possible through the threshing mechanism, so reducing the possibility of overloads. Equally, height sensing can hold the cutter bar close to the ground as the machine moves forward over undulating land. Next, the harvester's reel speed can be controlled in accordance with measured ground speed, to maintain the programmed relationship between the two speeds. Electronic control of threshing drum loading has been shown to be more rapidly responsive to changing field conditions than skilled operators – increasing work rate and simplifying the operator's task. Drum loading is monitored by sensing its rate of rotation: heavier loading reduces that rate. When it falls below or rises above the value preset by the operator the machine's ground speed is automatically decreased or increased to maintain the required feed rate. Automatic steering, based on sensors that can track the edge of a wheat crop, is the most recent addition to these control systems.

Massey-Ferguson's Daniavision in-cab monitoring and control unit signalled the move away from multiple dials and indicator lights to an integrated, computer-based display and programming unit. This unit provides alphanumeric information and bar chart displays which enable the operator to set up the machine for any commonly combined crop, at one of three moisture levels (dry, normal, moist), to call up monitored engine and shaft speed data, machine settings, current and total harvesting statistics, and to obtain information on servicing schedules and routines. Visible and audible warnings of malfunction or potential malfunction are given and the computer even shuts down the machine automatically in extreme circumstances, in order to forestall major damage.

3.4 Precision agriculture

There are numerous terms for the concept that has caused intense interest among researchers, manufacturers and farmers in many countries in the 1990s. Apart from the above section heading, it is possible to choose from precision farming, prescriptive agriculture, site-specific crop management, spatially variable field operations and spatially variable agricultural production systems (SVAPS) *inter alia*. This is a concept that originated in the early 1980s as researchers and manufacturers of electronic equipment for agricultural use speculated on the implications of microelectronics, robotics and satellites for farming at the start of the next millennium. The potential civil uses of the American Global Positioning System (GPS) were among the developments considered. Researchers began to explore varied facets of the subject from then on. At the same time the electronic instrumentation and control systems outlined in section 3.3 evolved to provide the necessary base for transfer of research ideas into practice. The concept then took a great stride forward when the GPS system became generally available for navigation and position fixing, albeit at a lower precision than that available to the military, but at moderate cost.

The Global Positioning System

The USA's GPS NAVSTAR has a constellation of 24 satellites, orbiting the earth in 11 hours and 48 minutes at an altitude of 20 200 km, in six orbits (four satellites per orbit) inclined at 55°. Their function as positioning references depends upon the measurement of time to the highest possible accuracy. To that end, as stated in section 1.1, they all carry a caesium clock. Their clocks are not periodically adjusted to solar time, therefore they deviate increasingly from that time. However, it is essential that they do not deviate from each other by more than a few nanoseconds, because time synchronisation is the basis of the system. In fact one of the immediate problems to be overcome was that relativistic effects greatly exceed the above limit: the satellites' clocks gain nearly 40 μs a day relative to earthbound clocks. This problem was overcome by setting their clock frequency lower than the ground-based system frequency by a small amount (10.229 999 995 45 MHz, compared with the system's 10.3 MHz). Ground control's task involves correction of these clocks and monitoring of the satellites' orbits (see Fig. 3.17).

The satellites transmit coded data at two L-band frequencies; both multiples of 10.23 MHz. L1 = 1575.42 MHz and L2 = 1227.60 MHz. These carriers are modulated by code sequences which are duplicated in the GPS receivers. When a receiver picks up a satellite transmission it adjusts the timing of its own code to that of the satellite. That provides the information required to determine the time delay between transmission and reception of the signal. Then the satellite's range is calculated by reference to the velocity of propagation of electromagnetic radiation (commonly referred to as the speed of light, as in section 3.2). In this context the most accurate value of the velocity is required, i.e. 299 792 458 m/s (in vacuum, which it mostly is en route to earth). The resulting value is known as the pseudorange, since there are corrections to be made. For example, the radiation is liable to refraction when passing through the ionosphere, with its layer of charged particles. This can be compensated by comparison between the L1 and L2 transmissions.

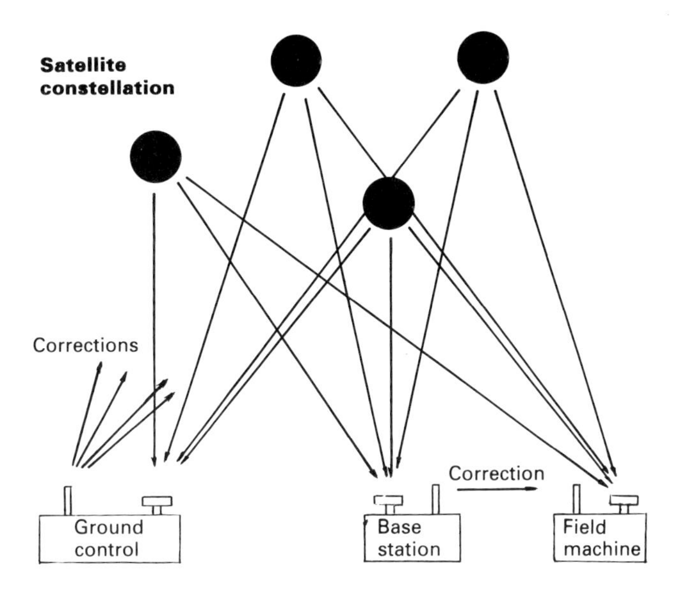

Figure 3.17 Global positioning system and differential global positioning system.

A range measurement on a single bearing cannot locate a position in either two or three dimensions, of course. Signals from at least three satellites are needed for a two-dimensional position fix and at least four for three dimensions. With 24 satellites in the above orbits three-dimensional location can be achieved and the receiver's coordinates on a fixed x,y,z (Cartesian) frame of reference can be determined. These coordinates are then transformed in the receiver to longitude, latitude and altitude by further reference to a model of the earth's ellipsoidal shape – the World Geodetic System (1984). Here again, some errors are bound to occur because the receiver has to interpret its position from a set of four pseudoranges, each with an error. The cumulative error is greater when the four satellites are in the same region of the sky, as seen from the ground. Therefore the receiver's suite of complex algorithms need to select the four most widely placed satellites. That can be done by reference to satellite almanac data.

However, when all possible corrections have been made there remains the matter of selective availability (SA) applied by the US Department of Defense. This distinguishes two sets of codes – the Precise Positioning Service (P code) and the Standard Positioning Service or Coarse Acquisition Code (C code). Civil users normally have access only to the latter, which is of lower precision, offering a resolution of about 100 m at 95% probability and 300 m at 99.99% probability. This is clearly inadequate for agricultural field work, therefore it is necessary to employ the differential GPS (DGPS) system (see Fig. 3.17) which improves resolution by a factor of 20. This uses two receivers, one of which is a base station, placed in a precisely known position. The other, the *mobile*, is placed at the field locations where the space coordinates are required. The base station computes its position via GPS then computes a correction derived from its actual position.

[Note: improved corrections are now available via an independent communications satellite.]

The correction can be applied to the mobile's information in real time, or later, depending on the requirement. If the latter is required to provide a continual correction of the mobile's field position, there is already a standard for the transfer of pseudorange correction data, which is implemented in many GPS receivers. This is the Radiotechnical Commission for Maritime Services format RTCM SC104 (1989). For field machines the data can be transmitted by an RS232 radio modem at UHF, as mentioned in section 2.2 (wireless transmission).

The rate of acquisition of positional data is dependent on the form of the receiver. A multichannel instrument, capable of simultaneous processing of data from several satellites, will be faster than a single-channel instrument with a multiplexer. For three-dimensional position fixing the multichannel form with at least four channels is the obvious choice. With this type of equipment position computing can be done at 1 s intervals, and the whole mobile system is readily portable (see Fig. 3.18(a)).

Site-specific treatments

Figure 3.18(a) shows weed mapping in progress, using the eye of the observer to register the positions and extent of patches of weeds as he walks a field, and the DGPS system to register his position each time he makes an entry in the palm-top PC that he is holding. The DGPS feeds the positional information into the PC via an RS232 link. This operation is part of the concept of patch spraying, based on evidence from agronomic studies from the mid 1970s onwards that many weed species occur – and persist – in patches of fields. It follows that if the position of those patches is known the application of herbicides (including pre-emergence treatments) could be made much more site-specific than in the past. If the species are identified, that increases the opportunity to make the treatment weed-specific. Alternatively, very precise mechanical forms of weed control might prove possible.

Weed mapping can be done by aerial survey in some circumstances, where weeds can be distinguished clearly from crops and soil. Aerial platforms such as captive balloons, radio-controlled model aircraft and light aircraft have been employed for this purpose. Field machines with cameras can also collect the required data. This process generally depends on differences in the spectral reflection of weeds, crops and soils. Unfortunately, the differences are frequently too slight for simple identification techniques. Even at crop level, distinctions of colour and/or shape are not readily convertible to automatic sensing. Research on the spectral reflectance characteristics of crops and weeds, together with advanced methods of image analysis, may yield means for automatic weed detection. Until then – or perhaps even then – the field walking and visual identification system is the most assured method of data collection.

An example of the end product is the map shown in Fig. 3.18(b). This is a

(a)

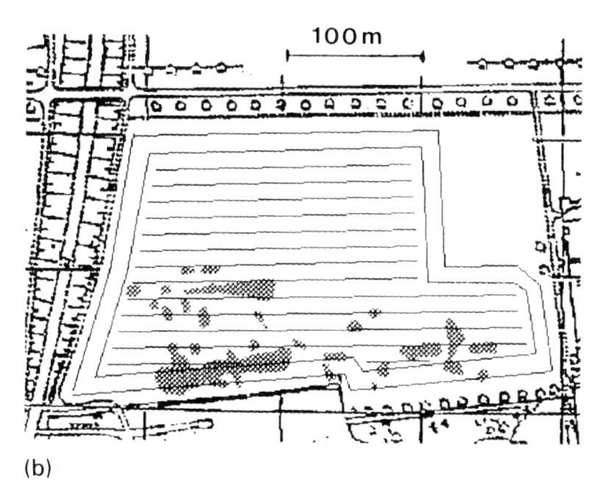

(b)

Figure 3.18 (a) Portable DGPS equipment for field recording; (b) computer map of weed patches in a cereal crop (Silsoe Research Institute).

composite, comprising a field map overlaid with the weed patch locations in a cereal crop and the tramlines which provide the guidance needed for the operator of the spraying machine. The map can be incorporated in a laptop PC, together with the sprayer's position at any time, as determined by the GPS, or by a combination of the tramline number and a ground speed sensor, which makes it possible to compute the distance travelled. A PC has been used to control a dual version of the spray control system shown in Fig. 3.11(b), i.e., with two metering cylinders and two corresponding boom manifolds in parallel. This arrangement made it possible to apply a base dose rate from one set of nozzles, supplemented at patch locations with a dose from the other set, or a choice of two separate applications rates, or a choice of herbicide. The input to the closed loop control of flow rate was achieved by monitoring the rotational speeds of the pumps on the metering cylinders, together with their associated pressure sensors. A PID control algorithm was employed. Control signals from the PC were transmitted via a serial data link.

A sprayer has in-built time delays, which can be several seconds. These have to be accommodated but must be minimised when it is used for patch spraying. For uniformity of response pipe lengths from the herbicide injection point to separate groups of nozzles must be of equal length and of minimum bore consistent with an acceptable pressure drop. The response time of the metering system is crucial, too, since it is required to react to sudden demands which are exacerbated when the whole boom is turned on at once. Control strategies to minimise the delay in achieving the required dose rate are needed.

Overall, research has shown that patch spraying can be controlled to a resolution of 2 m × 1 m (width × distance) or better.

Portable GPS equipment has also been used for site-specific surveys of soil nutrient status. This information, together with crop yield data, can provide further overlays on field maps, thereby creating a farm-scale GIS which can be added to year by year. It is widely recognised that information of this kind is essential to identify the ways in which the technology of precision agriculture can be most effective. In addition to patch sprayers this technology includes seed drills and fertiliser distributors with controllable discharge rates, already mentioned in section 3.3. Like the patch spraying system, these implements are controlled by a tractor-borne PC which can be supplied with site-specific data via a data transfer device such as a PCMCIA card.

Yield mapping

Figure 3.19(a) is claimed to be the world's first yield map (1985). It was generated by a Massey-Ferguson grain combine fitted with a prototype of the company's current mass flow yield meter (Fig. 3.16(b)). Since the GPS was not available at that time the field was divided into a 10 m square grid and when the combine entered a square the yield was recorded by hand. This clearly showed the con-

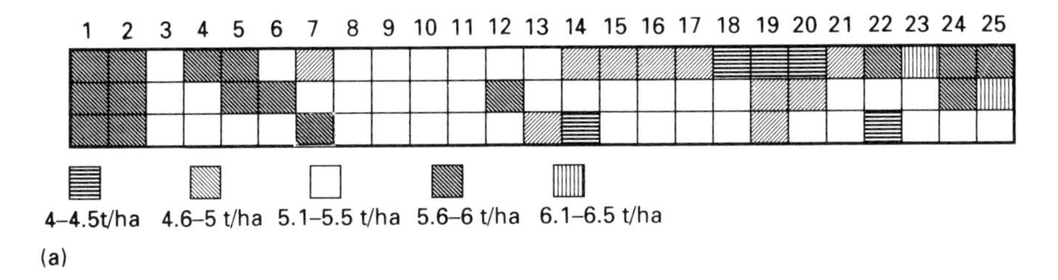

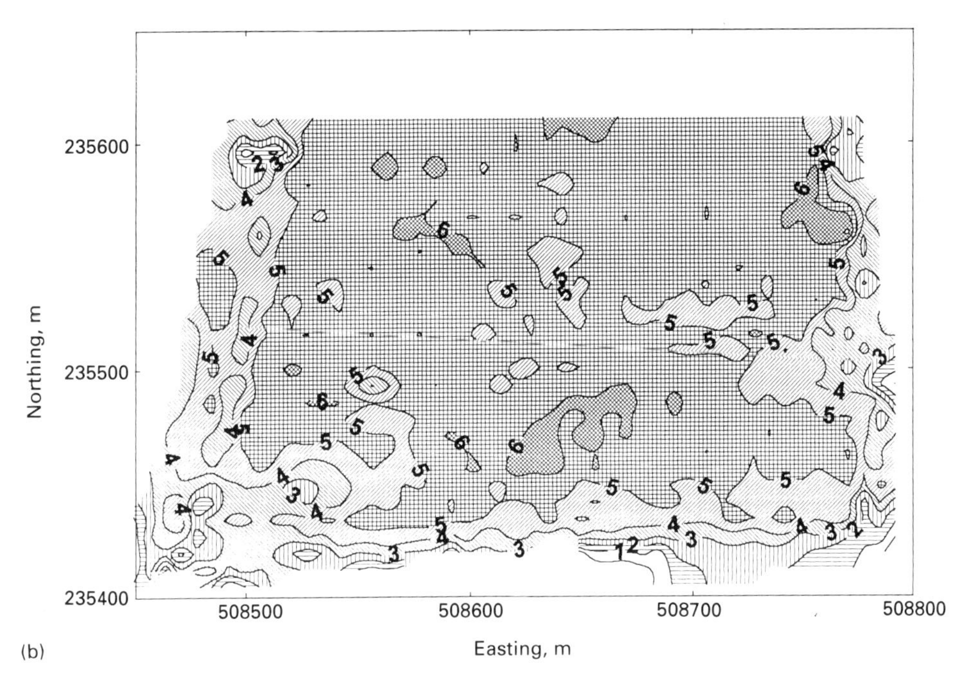

Figure 3.19 Yield maps (cereals): (a) 1985 (Massey-Ferguson); (b) 1994 winter barley, t/ha (Silsoe Research Institute).

siderable variation that can occur – and which is frequently unexpected. One of its descendants is shown in Fig. 3.19(b). It was derived from a combine fitted with a GPS receiver, working with a base station, and the output from the machine was transferred to the farm office by a smart card. The complex nature of the management options that such maps pose is self-evident.

Yield mapping is not restricted to cereals. Figure 3.20(a) shows a tractor/trailer combination collecting sugar beet from a six-row harvester. The tractor has a DGPS system (the antennae are visible on the cab roof) and a set of four electrical load cells is mounted on a subframe beneath the trailer body. The combined load cell output is filtered and smoothed to provide the cumulative data from which yield variation is derived. The combination is able to work with other root crops, onions and forage crops such as grass and maize. Figure 3.20(b) shows a yield map for potatoes.

(a)

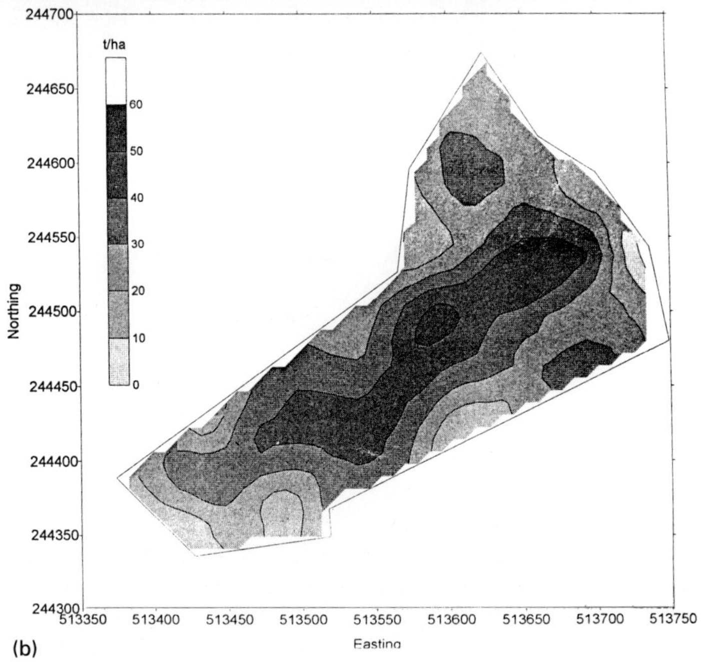

(b)

Figure 3.20 Yield mapping for root crops: (a) weighing trailer in use with sugar beet; (b) yield map for potatoes (Massey-Ferguson: Centre for Precision Farming, Cranfield University).

Field navigation

The ability of a field machine or implement to follow a prescribed track with specified precision is an obvious component of site-specific operations, in addition to the crucial manoeuvres described in section 3.3. That section contains examples of methods of steering implements relative to a tractor. Another example is automatic steering of a semi-mounted, multirow hoe, in order to maintain close lateral control at forward speeds higher than those attainable with manual steering alone. This can be done without the use of electronics by creation of a durable guideline slot in the soil, made by a suitably designed deep tine at drilling time.

Where it is necessary to navigate the machine itself, reference has been made already to the value of tramlines in a cereal crop as a guideline for the operator. The DGPS (civil version) provides a more general navigation aid, albeit with lower precision. However, advancing technology has led to the revival of interest in the development of vehicles with autonomous guidance systems, which were a feature of the driverless tractor concepts of the 1950s to 1970s. Those earlier developments included leader cable systems in which inductive sensors on the front of the tractor were employed to position it over or between buried sets of parallel cables, energised at low kHz frequency. This system achieved brief commercial status. Its control was good but the cables were vulnerable and the cost of laying them too high. In addition, they constrained the machine to work only in specific directions. Optical line of sight systems, based on the newly developed laser, found application in control of drainlaying machines, to achieve both the required line and gradient, but were otherwise limited to field guidance over limited areas. Mechanical and optical furrow following proved feasible, and headland turns could be made with the aid of on-board optical systems, cooperating with passive boundary reflectors, but these methods were not tested for the full range of field operations. Inertial systems did not achieve the required accuracy, nor did existing RF position-fixing techniques.

Therefore research has moved on to machine vision as the basis of the autonomous vehicle. This seems the most likely direction of progress, unless the full capability of the GPS is made available to civil users. In that connection, before access to P code was withdrawn American research workers reported that a DGPS combination of P code and carrier phase data could provide real-time navigation to a precision of ± 20 mm, with 0.5 s updates. Inevitably, this concept has not progressed in agriculture, but the application of machine vision has advanced, and an example of its use is given in the next chapter.

3.5 Remote sensing

Data on the spectral content of solar radiation reflected from the earth's surface are regularly provided by satellites orbiting at a height of about 700 km. Images cover a land area of nearly 200 km^2 and their pixel size is about 10 m. The reflectance spectra are split into wavebands from which *thematic maps* can be drawn. In the UK these are provided by the National Remote Sensing Centre. Maps of this type can distinguish individual crops by their reflectance signatures, as well as crop water content and biomass (yield). They need specialist interpretation because the reflected radiation is dependent on a variety of crop and atmospheric factors, but their application to classification of land use, soil classification and yield forecasting has given them considerable importance.

Other satellite systems can monitor large areas of forest or grassland for outbreaks of fire, which they can detect when only a few tens of metres in extent. They can also assess the risk of fire by monitoring risk factors such as surface temperatures, the dryness of vegetation and atmospheric water.

Surveys with airborne or spaceborne synthetic aperture radar (SAR) equipment have been employed to detect subterranean water supplies in arid regions. The SAR system collects radar echoes from a lateral band of the earth below, dependent on the beam width and the height of aircraft or satellite. As it moves it builds up the longitudinal elements of the radar image, correcting for the Doppler frequency shift of each echo, to synthesise the beam width (aperture) in that direction. Clouds and vegetation are transparent to the radar frequency (in the 900 MHz range). Echoes therefore return from the earth's surface and from underground discontinuities, as outlined in section 3.2 (GPR).

Ground-based radar is employed to monitor migration of insect pests. A compact, vertical looking beam, operating at 940 MHz, is the basis of equipment for long-term, routine monitoring developed at the UK's Natural Resources Institute. It can detect individual insects flying up to 1 km above the ground and can provide data on the insect's height of flight, mass, body shape, wingbeat frequency (large species only), ground speed and direction, and airspeed (i.e., after correction for wind velocity). When there are multiple targets the system can determine their volume density, height/density profile, area density profile and migration rate.

At even higher frequency – in the visible light range – the optical equivalent of radar (lidar) is being evaluated as a means to monitor the health of large areas of vegetation remotely. The laser impulses on which lidar is based cause plant fluorescence which can be detected by the echo receiver and spectrally analysed. This development has been stimulated by laboratory studies of laser-induced leaf fluorescence in the 400 to 800 nm waveband. Processing of the resulting images can reveal subtle changes in colour which relate to early signs of nutrient deficiency and other plant stresses before they can be detected by the eye. The method therefore has the potential to increase the precision with which pesticides, fertilisers and water are used. The development of ground-based, mobile systems, capable of remote sensing in the field, is possible. It has been reported that nutrient deficiencies in maize have been detected at distances up to 100 m.

3.6 Further information

Specific texts on meteorology, soil and soil water and crop production provide the scientific context for many of the measurement techniques covered in this chapter. Fuller details of those techniques can be found in general texts on instrumentation and control systems. Among the many relevant papers published in scientific journals the following (mainly reviews) amplify specific themes in the chapter and contain further references of value, including those on standards. The relevant sections in the chapter are indicated.

Section 3.2

Kano, Y., McClure, W.F. & Skaggs, R.W. (1985) A near infra-red reflectance soil moisture meter. *Transactions of the American Society of Agricultural Engineers*, **28**, 1852–5.

Whalley, W.R. & Stafford, J.V. (1992) Real time sensing of soil water content from mobile machinery: options for sensor design. *Computers and Electronics in Agriculture*, **7**, 269–84.

Whalley, W.R., Leeds-Harrison, P.B., Jay, P. & Hoefsloot, P. (1994) Time domain reflectometry and tensiometry combined in an integral soil moisture monitoring system. *Journal of Agricultural Engineering Research*, **59**, 133–40.

Section 3.3

Frost, A.R. (1990) A pesticide injection metering system for use on agricultural spraying machines. *Journal of Agricultural Engineering Research*, **46**, 55–70.

Howes, P., Law, D. & Dissanayake, D. (1986) The electronic governing of Diesel engines for the agricultural industry. *SAE Technical Paper Series 860146*. Society of Automotive Engineers, Inc, Warrendale, Pennsylvania.

Richardson, N.A., Lanning, R.L., Kopp, K.A. & Carnegie, E.J. (1982) True ground speed measurement techniques. *SAE Technical Paper Series 821058*. Society of Automotive Engineers, Inc, Warrendale, Pennsylvania.

Scarlett, A.J. (1993) Integration of tractor engine, transmission and implement depth controls: Part 2. Control systems. *Journal of Agricultural Engineering Research*, **54**, 89–112.

Sokol, D.G. (1985) Radar 11 – a micro-processor based true ground speed sensor. *ASAE Paper 85–1081*. American Society of Agricultural Engineers, St Joseph, Michigan.

Tinker, D.B. (1992) Integration of tractor engine, transmission and implement depth controls: Part 1. Transmissions. *Journal of Agricultural Engineering Research*, **54**, 1–27.

Yasin, M., Grisso, R.D. & Lackas, G.M. (1992) Non-contact system for measuring tillage depth. *Computers and Electronics in Agriculture*, **7**, 133–47. (Reviews ultrasonic and other sensors.)

Section 3.4

Auernhammer, H. (ed.) (1994) Special issue: global positioning systems in agriculture. *Computers and Electronics in Agriculture*, **11**, 1–96.

Stafford, J.V. (ed) (1996) Special issue: spatially variable field operations. *Computers and Electronics in Agriculture*, **14**, 99–254.

Tillett, N.D. (1991) Automatic guidance sensors for agricultural field machines: a review. *Journal of Agricultural Engineering Research*, **50**, 167–87.

Section 3.5

Bird, A.C. (1991) Principles of remote sensing: electromagnetic radiation, reflectance and emissivity. In: Belward, A.S. & Valenzuela, C.R. (eds) *Remote Sensing and Geographical Information Systems for Resource Management in Developing Countries*, pp. 1–15. Kluwer Academic, Dordrecht, The Netherlands.

Smith, A.D., Riley, J.R. & Gregory, R.D. (1993) A method for routine monitoring of the aerial migration of insects by using a vertical-looking radar. *Phil Trans Roy Soc Lond, B*, **340**, 393–404.

The WorldWideWeb is an increasingly fertile source of information on topics relevant to this chapter. Search for 'precision agriculture' and 'site specific agriculture' yields updates from several search engines. The SAE site can be visited via 'sae' to obtain the latest details on its standards, including J1939, which is based on CAN, version 2. Among the sources of scientific information, a search for 'nmr' yields a tutorial, with animation.

Chapter 4
Horticultural Crop Production

4.1 Introduction

This chapter is devoted to three aspects of horticulture, namely rowcrops, fruit growing and protected cropping. Since the dividing line between agriculture and horticulture is fuzzy in the context of field crops, some of the material appropriate to the chapter appeared in Chapter 3.

4.2 Field crops

Here, precision in field operations from drilling or planting to harvesting is the common requirement, in order to grow produce of as uniform size, shape and quality as possible. Among those operations one of the most important is assurance of the water supply required by all vegetable crops. Efficient use of water is increasingly necessary, therefore controlled irrigation is assuming greater importance.

Irrigation

Tensiometers, electrical moisture meters and evapotranspiration meters have been mentioned in section 3.2 as the basis for irrigation control, particularly in the horticultural sector. There, too, reference was made to computer control of pipeline distribution networks. Mobile wide-boom irrigators also play a substantial part in the application of water to vegetable crops in particular, and these have incorporated electronic control since the early 1980s. As an example, the control panel layout of a constant speed controller for a high pressure hose drum irrigator is shown in Fig. 4.1. Table 4.1(a) lists the quantities that its computer monitors and controls, while Table 4.1(b) lists the programmable constants (settings), with their allowable upper and lower limits and the factory adjusted default values.

The two control sensors are magnetically operated to ensure their environmental robustness. One is an on–off device, actuated by a stop-bar linkage which also selects neutral gear in the drum drive. The other is a rotary disc sensor in the drive itself, which provides a digital output that enables the computer to calculate

Figure 4.1 Hose drum irrigator: control panel (Briggs Irrigation).

the payed-out length of hose and the speed of reeling it in. These together with the timer and a water pressure sensor make up the inputs to the control system. The outputs are provided by the pulsed motors which operate the turbine bypass and shut-off valves. The former valve's setting during a run is modulated to adjust the speed of reeling in to its preset value. When the valve is fully closed the speed is at its maximum in the selected gear. The shut-off valve opens and closes the water to the irrigator. The hardware also includes a solar panel for recharging the system's battery when required.

When the START key is pressed, if the stop sensor is not activated (drive in neutral) the bypass valve cannot be closed, therefore the turbine does not run. The shut-off valve opens, but only briefly, to release excess pressure in the hoseline, making it safe to uncouple the line at a hydrant, if required. Otherwise, at START the bypass valve is closed to start the turbine, then the shut-off valve opens to begin irrigation. If pre- and post-irrigation periods have been programmed in (to avoid lower irrigation at the start and finish of a run), at the start the hose is reeled in a short distance ($\frac{1}{2}$ m) before the turbine is stopped and water is applied for the pre-irrigation phase. The computer calculates the time (in minutes) for this phase by multiplying the constant 1 (Table 4.1(b)) by the time for running 1 m at the programmed speed. Thereafter, the run continues at a constant set speed or at a preset sequence of different speeds over specified distances, to provide zoned irrigation. The speed can be changed at any time during a run, except when the motorised valves are operating. No keys are active during those periods.

At the end of the run the turbine is stopped by the stop sensor and if post-irrigation has been programmed the shut-off valve shuts down after that period, which is calculated in the same way as the pre-irrigation time. The run ends earlier if the automatic shut-down feature calls for it. This is activated if the machine irrigates at the same place for longer than the time set by constant 3 (default time 20 min). This could occur if the wrong gear had been chosen for the preset speed, or if

Table 4..1 Hose drum irrigator. (a) Monitoring and control functions; (b) programmable settings.

(a) KEY TO ACCESS DISPLAY

- Speed
- Total irrigation time
- Length of the pipe
- Pressure sensor
- Stop sensor
- Pre- and post-irrigation
- Timer, time to start
- Speed sensor
- Motor 1, bypass valve
- Motor 2, shut-off valve
- Battery voltage
- Charger on/off
- Actual speed
- Elapsed time and elapsed distance

(b)

Const no	Fact adj	Min value	Max value	Description
0	—	0	65h00	Timer
1	8	1	15	Pre-irrigation
2	8	1	15	Post-irrigation
3	20	0	99	Automatic shut down (supervision time)
4	2	4	4	1 English, 2 Danish, 3 German, 4 French
5	0	0	1	0 = Stop for high pressure, slow shut down 1 = Stop for low pressure, 1 long pulse and the motor runs in the opposite direction 2 = The motor for stopping is disconnected
6	0	0	15	Distance to post-irrigation (m)
7	—	0	1000	Distance (only for test) (m)
8	0	0	1000	Distance for bipper (0 = no bip)
9	100	—	—	Code to reach machine data

the water pressure was insufficient. If the *bipper* setting is *on* (non-zero) a 12 s radio or light alert is activated, to indicate that a preset unreeling distance has been reached, or that the stop sensor or automatic shut-down have been activated.

The operator can monitor the system via six display menus, called up by MENU and the UP/DOWN keys. Programming of speed is done by calling up MENU 1 (speed and time), which also warns when the battery voltage is low. The UP/DOWN keys are used to set the required speed. However when it is set at 11.1 m/s, the PROGram key calls up the constants in Table 4.1(b), which can then be changed as required. If the new constants are not *saved* by pressing the MENU key the program reverts to the original default values, which are retained in memory even if the battery is disconnected for a long period.

Irrigation dosing

Controlled dosing of irrigation water with concentrated additives provides another application of electronic closed-loop control, which can be employed in the field on mobile machines as well as in pipeline assemblies. The objective is normally to add the liquid concentrate in proportion to the amount of water, but on field machines the dosing rate may be made proportional to the speed of the machine. Additives that can be applied in these ways include not only fertiliser but also acids, to control pH and counter hardness in the water, and chlorine, added to water drawn from natural sources, to destroy pathogens. The control loop requires a means for monitoring water flow, a precise injector which can meter a specific volume of the additive into the water stream for each unit of water flow, a downstream sensor, to feed back the concentration of the resulting mixture, and a controller which regulates the frequency of operation of the injector. Some of these systems employ in-line water meters which produce electrical pulses at a rate proportional to the flow; others use paddle or turbine flow meters to produce a similar train of digital flow pulses.

An example of the latter type is shown in Fig. 4.2. Two injectors (2), powered by hydraulic pressure (5) and controlled by a correlator controller (1) from a paddle flow sensor (3), add concentrated fertiliser from the stock tank (6) to the water, downstream of the flow sensor (at 4). More than one injector has been employed here in order to obtain a high injection rate (say 1% after dilution). Further downstream, where mixing is essentially complete (7), an electrical conductivity (EC) sensor measures the concentration of electrically charged molecules (or ions) in the solution. These ions are created when mineral salts are

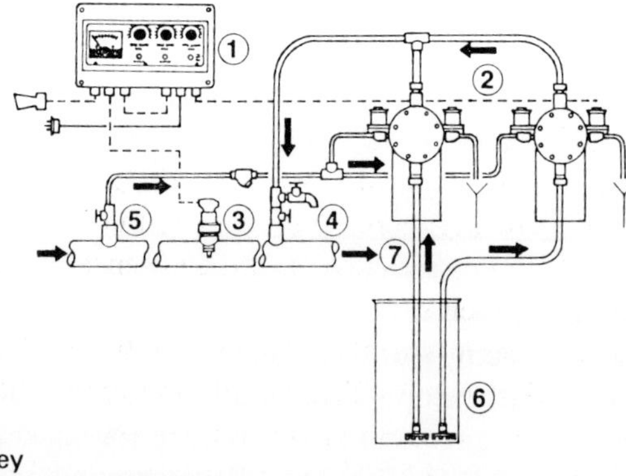

Key
1 Correlator controller/EC monitor
2 Fertilizer injectors(s)
3 Flow sensor
4 Fertilizer injection point
5 Hydraulic pressure
6 Fertilizer stock tank
7 EC sensor downstream to allow for in-line mixing

Figure 4.2 Dilution control (Rossell Fluid Control).

dissolved in water and their concentration is a measure of the overall concentration of salts in the solution (see section 1.16).

The correlator/controller allows the user to adjust the injector frequency – and hence the dilution ratio – over a wide range, very simply, by rotation of a control knob on its front panel. In operation, the user refers to tables which relate dilution to conductivity for a range of commercial liquid feeds. When the required dilution is established, this can be achieved by initiating water flow and increasing the injector frequency, until the conductivity meter on the correlator/controller indicates the required value. The control loop is then working at its set point and any subsequent variations in water flow or stock strength create an error which the controller will seek to eliminate by altering the injection frequency. It has adjustable high and low alarm settings, to cater for use of the wrong stock solution or for faults in the dilution system. It also warns when the stock tank is empty. Calibrations for particular solutions can be performed, if required, by preparation of a carefully diluted stock solution, checked with a hand-held conductivity meter. When calculating the settings required it is necessary to measure the conductivity of the water itself, which can be significant. The water's conductivity is added to that of the required solution, to establish the controller's set point.

If a controller of this type is required to maintain the pH of the water at a specific level of acidity, a flow-type pH sensor is employed. Chlorine sensing can be based on a different kind of sensor, known as an amperometric cell, coupled to an electronic amplifier/display unit by a screened cable. This makes use of the polarisation effect mentioned in section 1.16. The cell contains two electrodes of dissimilar metals, such as platinum and copper. When sample water from the main system passes through the assembly the electrodes create a current through the cell, with platinum the electrically positive electrode and copper the negative one. Hydrogen then appears at the platinum electrode, through electrolytic dissociation of the water, and reduces the current in proportion to the thickness of its layer. The presence of free chlorine in the water oxidises some of the hydrogen, in proportion to its concentration. In this way the concentration of chlorine is linearly related to the increase in current that it brings about. The associated circuit is used to zero the current in the absence of free chlorine in the water by *backing off* adjustments.

The range of measurement is about 0 to 150 mg/l $\pm 5\%$. The pH of the solution does not have appreciable effect over the range 4 to 7.5 pH, but there is a limit on water temperature (40°C) and a positive temperature coefficient of over $\pm 1\%/°C$ (e.g. a reading of 2 mg/l at 11°C will increase to 2.5 mg/l at 27°C). If the temperature of the water cannot be held sufficiently constant a temperature sensor, such as a thermistor, can be clamped to the inlet pipe of the sample cell and connected to the meter by a screened cable. This controls the amplification of the meter circuit in accordance with temperature. The cell must not be operated without water and at calibration it needs the unchlorinated water to flow through it for 8 hours before the zero setting is adjusted. This gives the hydrogen ion layer time to stabilise. Then the gain (amplification) of the amplifier can be adjusted to give the required span of the meter, by chlorinating the water to the required level. Again, this requires several hours of stabilisation. The measurement is always prone to fluctuations – changes of water speed over the electrodes are partly responsible – therefore the amplifier employs an averaging circuit. Swirling through the cell helps to reduce variation, too, by agitating glass balls in the assembly, which clean the electrode surfaces. After initial calibration, regular checks against laboratory determination of the free chlorine in the water are recommended for the maintenance of accuracy. Also, particularly where the untreated water is heavily polluted or a wide range of dose rates is required, the cell needs to be regularly flushed with a detergent solution, followed by clean water.

In hard water areas, the speed with which chlorine kills pathogens may be much reduced, compared with neutral or acidic water. In these circumstances chlorine and pH control systems may be combined.

Spraying

In section 3.2 it was noted that meteorological stations can be used to control irrigation and to predict the onset of crop diseases, as a means to schedule crop spraying. The conditions for the onset of hop downy mildew were among the first to be predicted in that way. The risk index (I) was calculated on a scale 0–9 from readings of the rainfall and surface wetness sensors over the preceding 48 hours, together with the average air temperature during the wet part of that period; I = 0 when rainfall = 0 and the average temperature was less than 8°C.

Apart from pests and diseases, avoidance of competition between plants is especially important in an industry aiming for consistent, high quality. In the 1970s that led to the development of rowcrop thinners with a measure of artificial intelligence. Given weed-free soil they were able to detect the position of seedlings in a row by electrical contact or by non-contacting optical sensors which could distinguish the seedlings from soil by their different spectral reflectance. On that basis they were able to decide which plants to remove in order to produce the nearest possible approximation to the desired, uniform stand. However, precision drills and pre-treated seeds produced more uniform emergence and the demand for such thinners disappeared. That still left the main competition – weeds – to be dealt with, using mechanical or chemical means.

The colour video camera (section 1.17), coupled with the development of computer hardware and software, has now brought machine vision and artificial intelligence to bear on the weed problem. The autonomous vehicle developed at Silsoe Research Institute demonstrates the techniques involved in precise location of rowcrop plants and of weeds, both inter-row and intra-row. The experimental machine is shown in Fig. 4.3(a). A forward- and downward-looking video camera captures 10 images/s of the ground ahead of the machine, taking in several rows.

Figure 4.3 Autonomous vehicle: (a) experimental vehicle; (b) processed image, showing calculated row structure (Silsoe Research Institute).

The images are processed to find the row structure and to distinguish the crop plants from weeds and soil.

Figure 4.3(b) shows a monochrome version of a processed infra-red image taken by the camera, showing part of three rows of cauliflower plants. The left-hand row has a conspicuous gap and the plants are interspersed with many small weeds. The field of view is somewhat less than 2 m^2. That provides the detail necessary to distinguish the weeds but only very limited information on the direction of the rows, since no more than four plants can be seen in any frame. Nevertheless the controller needs reliable information on the row structure in order to steer the vehicle precisely along the rows. This conflict has been resolved by a calculation known as the Hough transform. Details of the transform and its application to this problem are provided by a reference at the end of this chapter. In essence it builds up evidence for the most likely (peak) location of the row structure, as shown by the lines in Fig. 4.3b, taking into account geometric information such as the height and inclination of the camera. The transform has to be implemented 10 times/s in order to allow real-time control of the vehicle's steering.

The Hough peak is not calculated entirely afresh at each frame but is *tracked* by reference to its previous value and to the forward movement of the vehicle, as measured by ground wheel sensors and other inputs. This process, together with the fusing of information from three rows at once, makes the system sufficiently robust to accommodate misplaced or missing plants. Randomly located inter-row weeds are disregarded by the transform.

The vehicle has operated at speeds up to 1.5 m/s, with a steering accuracy of ±20 mm. It has been used initially for spot applications of herbicide to identified weeds, via the front-mounted spray bar that can be seen in Fig. 4.3(a). To do this the controller uses the incoming frames to build up a local map of the area between the image field of view and the array of nozzles on the bar. Then, at the correct time, the appropriate nozzles are activated, to target the identified weeds. This is only one of the many spot treatments of plants and soil that are envisaged.

Whether an intelligent machine of this kind will find application for harvesting field vegetables remains to be seen. Selective harvesting could be its most likely niche. Over many years there have been numerous investigations of electronic control applied to both selective and once-over harvesting of these crops. Selective methods have been based on colour, size, height and density – the last mentioned dependent on the plant's attenuation of X-radiation. Computer-based machine vision now has the capability for real-time selection, based on external features, as the autonomous vehicle has demonstrated. Use of X-ray equipment in the field is not out of the question, as was demonstrated by the former Scottish Institute of Agricultural Engineering (SIAE) in the 1970s.

The SIAE's X-ray separator of potatoes from stones and clods was one of the earliest practical applications of electronics to field harvesting. It was based on the principle that the transmission of low energy (*soft*) X-radiation can be used to discriminate between objects of similar size but different atomic constitution, such as crop material on the one hand and stones or clods on the other. The crop material is composed largely of hydrogen, carbon and oxygen, all of which are of lower atomic number than silicon, present in this context as the main constituent of stones and clods. The differential absorption between silicon and the other elements named increases with the softness of the X-radiation but there is a lower limit in practice, because the absorption of X-rays and gamma-rays by matter is exponential in form. Mathematically, if I_0 is the intensity of radiation falling on an object composed of, say, four materials and with a thickness x, the transmitted radiation is given by:

$$I = I_0 \exp.(-(\mu_1 \rho_1 r_1 \ldots + \mu_4 \rho_4 r_4)x)$$

where μ is the mass absorption coefficient for a specific material and radiation energy, ρ is the density of the material and r is the proportion of that material in the radiated object. This equation could represent absorption by a potato covered with soil. The characteristic of μ is that it increases steeply at lower energies for all elements and the reduction of the radiation by absorption (I/I_0) then becomes correspondingly large. Therefore if the emergent beam intensity, I, is to be big enough to detect conveniently, without the use of a very powerful source (i.e. I_0 very large), it is necessary to pick an X-ray energy of just sufficient softness to achieve the required discrimination. In the present instance this is the ability to distinguish large potatoes from the smallest stones or clods in a mixture of the three. The use of an X-ray tube with a target (anode) potential of 30 to 40 kV has been found to meet this specification.

The essence of the X-ray sorter for a potato harvester is shown diagrammatically in Fig. 4.4. The X-ray tube radiates horizontal beams through a set of windows in its collimator assembly. The beams irradiate a corresponding array of X-ray sensors. The harvested potatoes, mixed with the stones and clods that survive pre-cleaning, are fed onto an automatically levelled conveyor and reduced to a single layer by rotating rollers above the conveyor (not shown in the

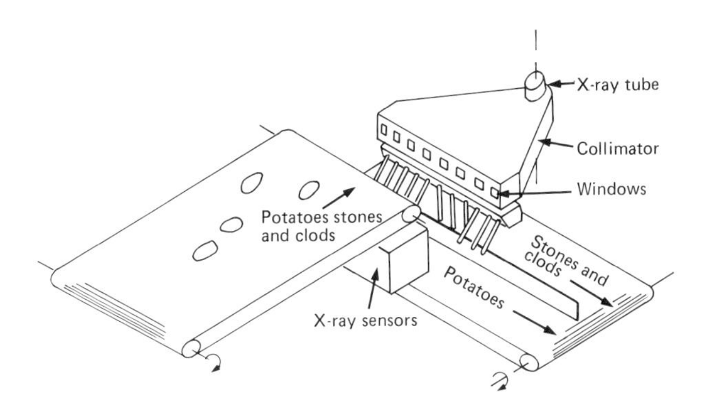

Figure 4.4 X-ray sorter for a potato harvester (Scottish Institute of Agricultural Engineering).

diagram). The objects all fall through the array of X-ray beams and are classified by the signal processing circuits which follow the sensors and which control pneumatically operated fingers. These are depressed to allow stones and clods to continue their downward fall, but they remain in their normal angled position to deflect potatoes into a separate discharge channel.

The original system was designed to segregate a minimum of 20 objects per second. Its X-ray sensors were scintillator crystals whose light output was then detected by sensitive photomultiplier tubes (see section 1.8). The photomultiplier current pulse represented the value of I in the preceding equation and this was compared with a reference pulse in an operational amplifier circuit. If the measured pulse was below the reference level the presence of a stone or clod was deduced and the circuit then operated the appropriate finger or fingers after a preset delay.

This system was capable of distinguishing between a 150 mm thick potato and a 25 mm thick piece of sandstone. In trials it separated potatoes from stones and clods with only a few per cent error, mainly among objects less than 50 mm in size. It is worthy of note that many of the development problems were centred on the means of presenting a steady flow of objects to the X-ray system. This is characteristic of automatic grading (see Chapter 5).

Two other sensing methods that provide information on conditions below the surface of a scene have been employed to find objects largely obscured by a leaf canopy. American workers have shown that standard, commercial ultrasonic transmitter/receivers – working at 50 kHz and pulsed at 10/s – can generate echoes from hidden cantaloupe melons, as well as soil, to provide real-time information for a harvester. The technique involves analogue filtering of the return echoes, then differentiating to produce sharp pulses corresponding to the acoustic discontinuities produced by the cantaloupe and soil surfaces. In this application it is necessary to make a correction for air temperature (see Table 3.2).

The second method, explored by Israeli workers more recently, is the use of close-range radar to detect the ground surface beneath a canopy. They used downward looking continuous wave, frequency modulated radar (FM-CW) at heights above the ground between 0.75 and 1.5 m. The carrier frequency was modulated between 8 GHz and 12 GHz (X-band) by a triangular waveform (Fig. 4.5). The return signal is displaced in time in proportion to the range of the reflecting surface and that creates a difference in frequency between the transmitted and return signals at a given time. The frequency difference, Δf, is constant over much of the modulation cycle (frequency fm). By simple geometry, Δf is related to the time delay, Δt, by $\Delta f = 2Bf_m\Delta t$, where $B = f_{max} - f_{min}$. The range $R = \frac{1}{2}c\Delta t$, where c is the speed of light, therefore:

$$R = \frac{c\Delta f}{4Bf_m}$$

This simple analysis assumes that there is no Doppler frequency shift between the outgoing and incoming radiation, i.e., that there is little or no relative movement

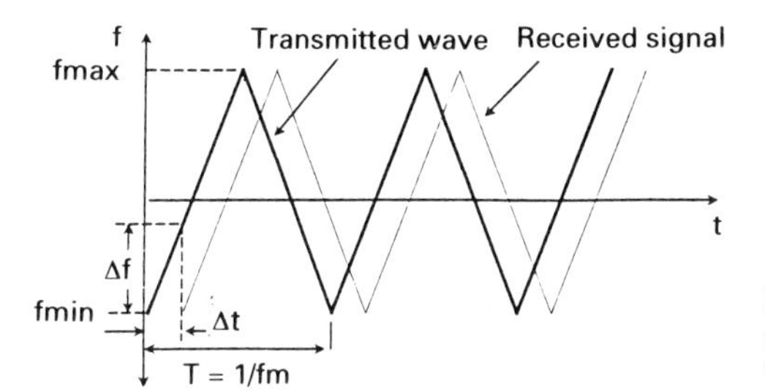

Figure 4.5 Height measurement, using FM-CW radar (Israel Institute of Technology).

between the radar antenna and the soil. Therefore the radar beam must be essentially vertical to the soil surface to meet that condition.

Investigation of this system showed that foliage produced many echoes, but that the underlying contours of the ground could be discerned. However, the algorithms required to extract the required signal from the noise had still to be developed.

4.3 Fruit crops

These more permanent crops require regular ground treatments to control grass or weeds – a task that is normally performed manually but which was one of the applications of the leader cable system (section 3.4) of the 1970s, because the tractor and implement could be guided round the orchard by a suitably laid cable. Another regular task – orchard spraying – does not deploy the sensors and actuators used in field spraying. However, electrostatically charged sprays have been used for spraying bush plantations in some areas. The charged spray droplets set up an opposing charge on the plants by electrostatic induction. This has the effect of drawing the spray to the surfaces of the plants, against the forces of gravity and airflow, and it will deposit on the underside of leaves as well as their upper surfaces, to provide improved protection against pests and diseases. Although this is not strictly instrumentation, it is a form of electrical control.

Intermittent spraying of fruit trees with water at the flower-bud stage is common, to avert damage to the buds in frost conditions by covering them with a thin, protective film of ice. Thermistor-based, simulated buds have been developed to control the application of water by sprinklers when it is required. This has been shown to require substantially less water than simple on–off regimes.

Later in the season, a research tool has found some application in horticulture and forest management. That is the sap flow sensor for automatic monitoring of trees' water use. The sensor is in the form of a three-spear probe. The tops of the collinear spears are pushed into the tree to make contact with the sap. A heat pulse is then generated at intervals by the centre spear and the outer spears register the resultant upstream and downstream temperature profile. Their differential output yields a reliable measurement of sap flow, in real time. An associated data processor can be used with multiple sensors to provide better sampling statistics.

In contrast to these isolated developments, since the early 1980s there has been widespread research aimed at selective harvesting of tree fruit by robot machines. Although some of this work has been for apple harvesting, much of it has concerned citrus fruit, for a variety of economic reasons. Citrus fruit also provide the advantage that their colour change at ripening is generally well marked. The essential requirements for these systems are:

- A means to identify the fruit that are ready for harvesting.
- A means to detect them when they are partly obscured by leaves, branches and other fruit.
- A picking arm able to reach them and to approach them gently, even when their range is not known exactly.
- A picking mechanism able to detach a fruit without damage to it and the tree.
- A means to deposit the picked fruit into a bulk handler without damaging them.
- A system able to compete economically with manual pickers.

An advanced concept from the early 1980s is shown in Fig. 4.6. This employs a single optical sensor to view each tree in turn from the nearest inter-row position. The sensor is a 256 × 256 array of photodiodes, connected in groups of 8 × 8, to divide the image into 32 × 32 subzones, as shown in Fig. 4.6(a). The four quarters

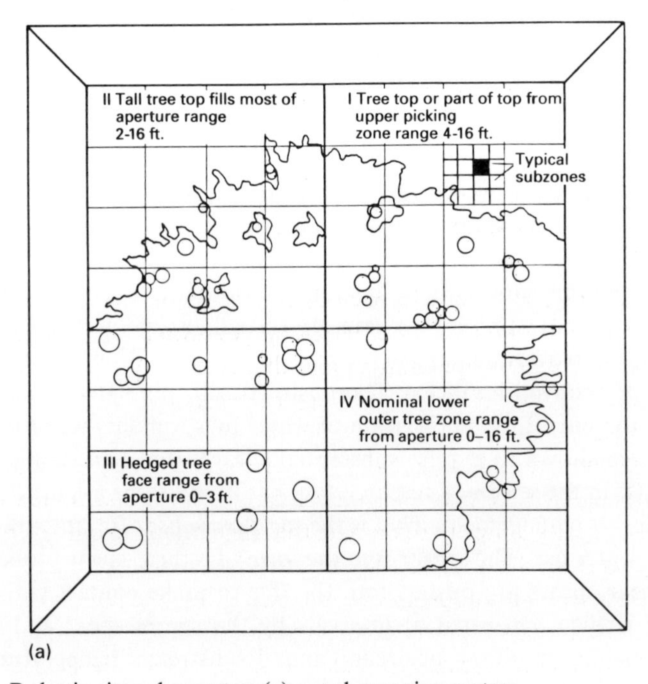

Figure 4.6 Robotic citrus harvester: (a) zonal scanning system.

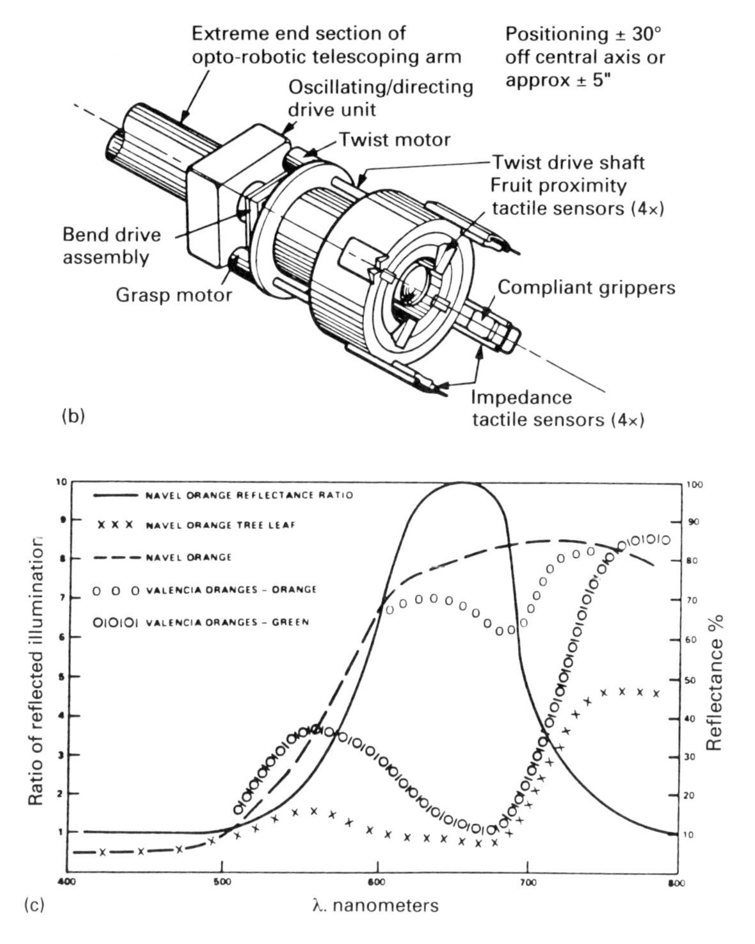

Figure 4.6 (cont.) Robotic citrus harvester: (b) severance module; (c) reflectance spectra (Martin Marietta Corp).

of the figure also show typical distributions of oranges in the field of view in different tree zones and at different ranges. Maximum contrast between the light received from the ripe oranges and that from leaves and unripe fruit is essential for identification of the target fruit. Therefore the sensor views the scene through an optical filter which transmits in the range 600 to 700 nm. The harvester must be operated at night to avoid unwanted reflectance effects due to sunlight (which has the advantage that the fruit are picked in the cool state). It therefore carries high-intensity lamps to illuminate the scene briefly, from several angles, while the outputs from the photodiodes are stored in computer memory. Each subzone in which a ripe orange has been detected is then visited in turn by an extended picking arm, which starts in the uppermost zone and works its way downwards.

The telescopic arm terminates in a severance module (Fig. 4.6(b)) which oscillates about the arm's axis and which has its own optical scanning system, sensitive to the 600 to 700 nm light range. When it locates a ripe orange it progressively slows the extension of the arm from its original speed of 4 m/s and directs the arm to the fruit. When the tactile sensors register contact the elastomer-tipped grippers close around the fruit. Then the module twists, bends and retracts sharply, simulating the action of human pickers, to remove the fruit from its stem. In order to harvest the oranges undamaged a secondary arm, below the severance module, collects them in a tray, from which they are transferred to the bulk bin.

The colour sensing is based on the fruit/leaf spectral reflectance ratios in the 600–700 nm band, for reasons that can be seen in Fig. 4.6(c). This well known relationship between leaf, mature and immature fruit spectra has been the basis for many sorting processes. Nevertheless, there is continuing research into forms of machine vision which will make it possible to perform these operations in natural daylight, despite its variable quality and its changing effect on shadows. This involves the complexities of colour analysis, which have to take account of the spectral composition of light incident on an object, the energy content (brightness) of that light, its saturation (its paleness or strength) and the optical properties of the reflecting object. The object's spectral reflectance is only one factor; parts of its surface may be glossy, others matt. The glossy parts will reflect specularly (like a mirror), while the matt (or body) reflectance is less dependent on the incident angle of the light.

Visual perception of colour has been classified by the CIE (Commission Internationale pour l'Eclairage), or ICI in its Anglicised form, which devised the chromaticity diagram shown in Fig. 4.7. In the diagram x and y are two of the three tri-chromatic coefficients defined by CIE (1931). These are derived from its X,Y,Z colour system, in which the three quantities represent wavebands in the visible spectrum, broadly centred in the red, green and blue ranges, respectively. The Y distribution approximates the eye's response, so it is a measure of brightness (or intensity). All colours can be resolved into a mixture of the three components. The definitions of x and y are $x = X/(X + Y + Z)$ and $y = Y/X + Y + Z)$. Then $z = Z/(X + Y + Z)$, from which $z = 1 - (x + y)$. Therefore a two-dimensional graph can define the visual intensity, hue (predominant wavelength) and saturation (or chroma) of a colour. The point E, with its x, y coordinates $\frac{1}{3}$, $\frac{1}{3}$, is where $x = y = z$, i.e. where the eye perceives white or grey. The hue of a colour is determined by the point at which a line from E, through the x, y coordinates of the colour, reaches the boundary. The wavelength associated with the boundary at that point is the hue. The colour's saturation is the fractional distance of the x, y point along the same line. Intensity, hue, saturation (IHS) analysis can be made in this way.

A later CIE standard, CIE 1976, employs the principal components L*,a*,b*, representing luminosity, green/red and blue/yellow, respectively. Luminosity is perceived brightness, taking into account the eye's response to light energy at different wavelengths, and akin to Y above. The long established Munsell colour system employs the parameters HCV, where H is hue, C is chroma and V is *value*

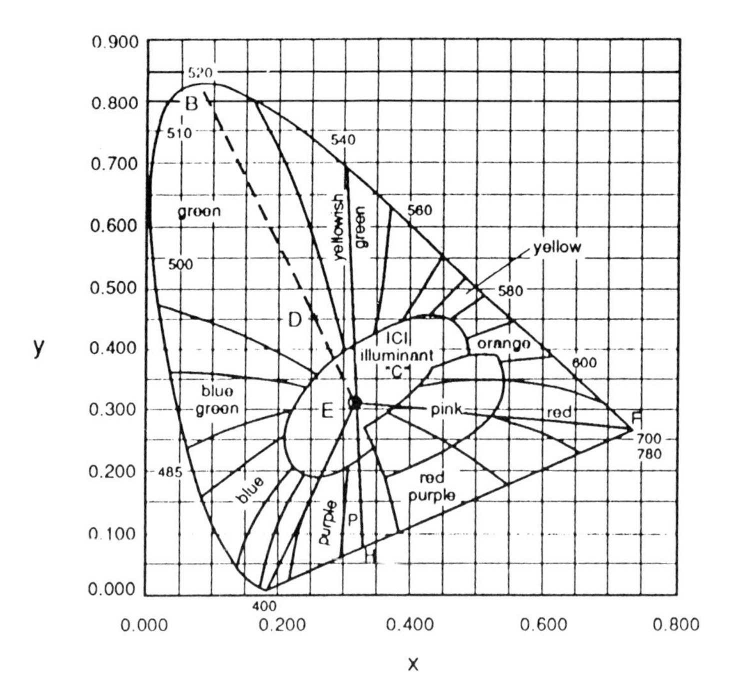

Figure 4.7 CEI chromaticity diagram.

(brightness). Here the RGB (red, green, blue) colour components of contemporary video signals provide a convenient starting point for image analysis, despite the fact that the CCD camera does not have the range of the human eye. Therefore most imaging methods for fruit detection now start with the camera's colour frames. Search algorithms generally look for clusters of pixels with the required RGB coordinates, although a variant method is to classify objects by reference to their surface (specular) and body (matt) colours, using polar spatial coordinates (r, θ, φ) instead of Cartesian coordinates (x, y, z). Whichever method is employed, once the (fuzzy) differences in the colour of the fruit, leaves and background have been established neural networks have to be trained to recognise the relevant clusters in the images. In that way even largely occluded fruit can be recognised. Further algorithms are required to resolve clusters of fruit into individuals.

These search techniques are exemplified in Fig. 4.8. Figure 4.8(a) shows a colour cluster that identifies an orange. Once the image has been colour segmented on that basis, then thresholded to remove background detail, the computer employs edge-fitting techniques to construct an outline of the fruit – extrapolating from a partial contour, if necessary. Figure 4.8(b) shows a computer's resolution of an image of three overlapping oranges.

Search techniques such as these can correctly identify between 95% and 100% of fruits that are more than 50% visible, and at an average rate of 0.3 s per fruit.

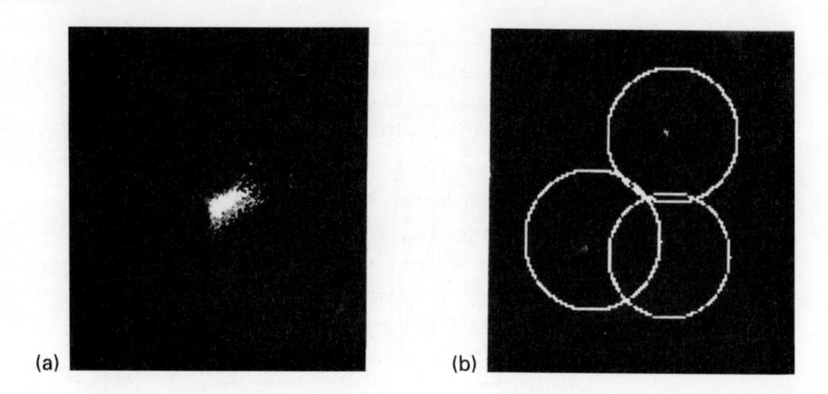

Figure 4.8 Image analysis: (a) colour cluster, identifying a fruit; (b) edge detection (University College, London).

That compares well with human capabilities. However, the rate at which the picking mechanisms can operate without damaging the produce, in comparison with manual harvesting, will determine the viability of these systems.

4.4 Protected crops

The applications of instrumentation and control systems in this sector are mainly restricted to the production of high value crops, including flowers and other ornamental plants, grown in heated greenhouses. There are also significant applications in the large-scale production of mushrooms.

Greenhouses

In northern European latitudes the main requirement for greenhouse construction is normally to maximise the use of available daylight, reserving the use of artificial illumination for those times when the level of naturally occurring, photosynthetically active radiation is inadequate. This has led to the design of the familiar glasshouse, with a minimal aluminium alloy framework, of sufficient strength to withstand normal (say, 50-year average) wind loading. A valuable property of glass is that it readily allows solar radiation into the house but its transmission of radiation is much lower at 10 μm wavelength, which is equivalent to radiation from a body at a temperature of 300 K, or 30°C. Therefore it traps much of the solar energy that has gone into warming the interior of the house. When solar energy is in excess of the immediate requirements, motorised ventilation panels or fans are needed to prevent sharp rises in air temperature. Fan ventilation can incorporate evaporative cooling of the incoming air by water spray or pads, but this is much more common in hotter regions. When solar radiation is insufficient, heat energy must be supplied from one source or another, and this can be responsible for over 40% of production costs. Despite the thermal properties of glass at longer wavelengths the house can cool quite rapidly, at a rate dependent on sky and wind conditions. At night, with a clear sky, the house will radiate energy to a near zero (K) heat sink. Double-skin construction of the envelope, with an insulating air space, can reduce this loss but only at the greater expense of light loss.

Overall, in order to maintain the close and uniform temperature regions that are needed for efficient crop production in conditions where heating and ventilation requirements are liable to rapid change, the temperature control system needs to be both sensitive and responsive. The system also has to control RH, which can create a conflict between RH and temperature requirements. For these reasons, glasshouse growers adopted electronic control ahead of most of the agricultural and horticultural industry. They have also led the way in making use of the computer for supervisory monitoring and control (i.e., SCADA systems) and for modelling their production systems.

The glasshouse also has its own range of control equipment for water application and nutrient feeding of plant roots, both of which employ sensors and control units adapted to those applications.

The aerial environment

Automatic monitoring of this environment calls for external meteorological inputs. In the glasshouse context the wind speed and direction must be known at about roof height (4 to 5 m) because of their interaction with the house's ventilation system. The strength of the wind has an effect on ventilation rate when the vents are not closed, of course, but apart from that the wind forces on the structure must be considered. These can damage open or partly open ridge ventilators and, in extreme conditions, can damage the whole structure, as already indicated. Heavy rainfall entering through open vents can damage the crop, therefore a rain sensor is another necessity. Although not as essential, a solarimeter at mast height is a valuable addition to a weather station, since its measurement of solar radiation can be employed in three ways. It can be used purely to provide information for the site manager's analyses, its output can be used to calculate crop water requirements, with the aid of published formulae, and it can be used in the control of carbon dioxide injection.

External air temperature is another quantity that should be monitored. This is measured within an aspirated screen (at about 1 m above ground level), to avoid exposing the sensor to solar radiation. *Inter alia*, this provides information on the temperature *lift* produced by the glasshouse heating system and that relates to the heat energy input. A second solarimeter, with a surrounding shade ring to prevent direct sunlight from falling on its sensor, can also be incorporated in the weather station. Comparison of the outputs from the two solarimeters provides the grower with information on the relative proportions of direct and indirect (diffuse) solar radiation falling on the glasshouse. This has an effect on the light transmission of the structure and hence on crop performance.

Most of the meteorological sensors have been described in Chapters 1 and 3 but the solarimeter merits special attention here. Its sensor is normally in the form of a set of thermocouple junctions in series, manufactured from alternate lengths of thermoelectrically dissimilar wires, such as copper and nickel alloy (Table 1.2). These are arranged radially on a flat surface to form two concentric rings of junctions. The central ring is covered with a specially blackened disc, which acts as the sensor, absorbing any solar radiation falling on it and being heated by it. The underlying junctions therefore become the *hot* junctions of the set of thermocouples. The outer ring is shielded from radiation and forms the set of cold junctions, which takes up the temperature of the solarimeter's body. The summated output voltage of the thermocouples, resulting from the temperature differential, is a measure of the solar radiation. A complete solarimeter assembly is shown in Fig. 4.9.

Figure 4.9 Solarimeter.

The blackened sensor disc, surrounded by a reflecting white ring, is mounted in the central plane of an accurately formed hemispherical cover (usually of glass). This allows radiation to reach the sensor from all directions above the horizontal plane, without degrading the uniformity of its directional (cosine) response. The body of the instrument is also shielded by the external white disc and the whole assembly is fitted on levelling screws, to ensure that the plane of the sensor is horizontal. A spirit level is incorporated for this purpose. The time constant of this instrument is made short (up to 30 s, depending on type) by reducing the mass of the element to a minimum. Well designed and maintained solarimeters are stable and they cover a wide range of radiation levels, but their output can change by 3 or 4% due to ambient temperature changes and their calibration should be checked occasionally (say, at 2- or 3-year intervals) by a specialist laboratory with the necessary calibration equipment.

Turning now to the uses of the other sensors in the environmental control system, the wind speed and direction measurements modify the action of the ridge ventilation control shown schematically in Fig. 4.10(a). This is interlocked with the heating system to avoid competition between the two. When actuated, the ventilators normally respond to a $P + I$ control algorithm individually tuned to the glasshouse characteristics. To avoid wind stresses on the structure though, the leeward ventilators open first – possibly by 50% of their total movement – before the windward vents start to move. Thereafter they will move in unison. The controller therefore has to know which is the leeward side at any moment. This information is derived from the wind vane's output, together with the known direction of the longitudinal axis of the house (E/W or N/S). To avoid unnecessary wear and tear on the ventilator drive (M) – a longitudinal shaft with rack-and-pinion aperture adjustment – reversal of the leeward/windward roles of the vents does not follow the indications of the vane at all times. Reversal only takes place when the necessary change in wind direction has been sustained for a preset time and the wind speed is above a preset threshold value (say, 0.5 m/s). However,

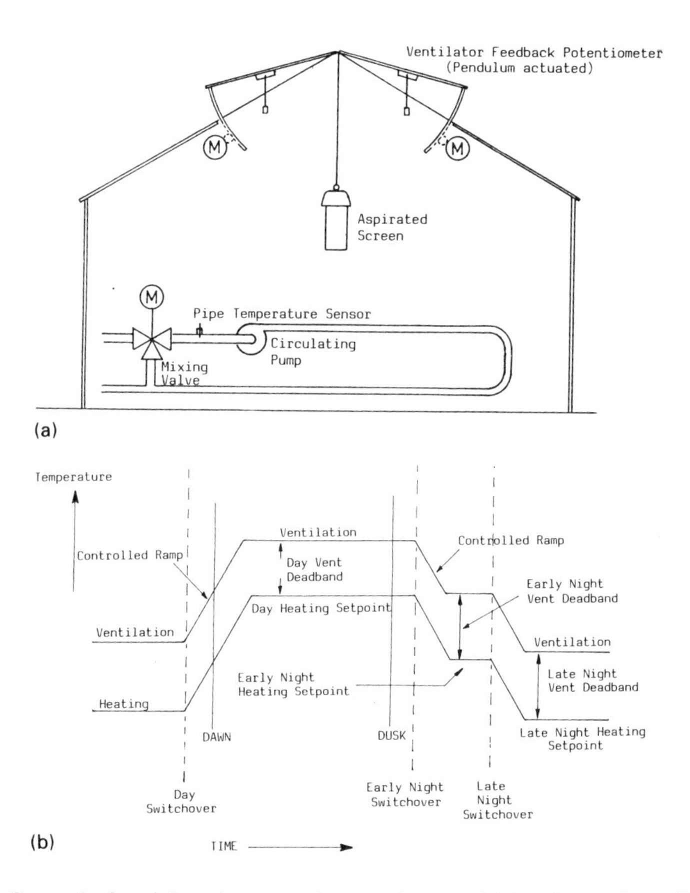

Figure 4.10 Control of aerial environment in greenhouses: (a) heating and ventilation system; (b) heating and ventilation profiles (Victor Automation Systems).

a fresh to strong wind (say over 10 m/s) will cause the windward vents to close, followed by the leeward ones, to avoid the potentially damaging structural stresses mentioned.

Inside the house, under normal conditions the ventilators operate to prevent the air temperature and RH from rising above their set points, in a closed-loop system, with the control algorithm and wind or rain overrides outlined above. The feedback sensor is usually a simple potentiometer device, such as the pendulum unit shown in Fig. 4.10(a). The slider is linked mechanically to the pendulum bob in this case. Its output at the fully open and full-closed positions of the vents is memorised by the controller, together with the time taken by them to travel between these extremes, under the control of their single-speed motors. This allows the control system to apply other constraints to the response of the ventilation system, as a means to reduce wear and tear. For example, the user can define a deadband (differential) within which an air temperature error produces no movement of the ventilator. Second, movement can be inhibited if the calculated travel time needed to cancel the error is less than a preset amount, say 0.5 s.

Figure 4.10(a) also shows a circulatory hot water heating system, which is the most common form of heating. Heating control is achieved by adjusting the temperature of the water, via the motorised mixing valve, which is opened or closed in response to the heat demand by a $P + I$ algorithm. As in the case of the ventilator control, hunting is avoided – and wear and tear on the valve minimised – by the introduction of a deadband, selected by the user. A further constraint can be applied by reference to the output of the pipe temperature sensor. This is employed to prevent the water temperature from reaching the level at which it could be hazardous to plants close to the pipe circuit.

Although hot-air heating from plastic ducts and radiant heating from the roof are found in some houses, all heating and ventilation systems respond to the dictates of the air temperature sensor. This is mounted in the aspirated screen (Fig. 4.10(a)) which is shown in the normal central position that it occupies in the glasshouse, or in any separately controlled compartment of a multi-compartmented house. A single sensor is adequate in the glasshouse context, because environmental conditions in the house are normally uniform to ± 1°C. One design of the screen is shown in Fig. 1.6(b). This enables the controller to hold the air temperature in the house to within ± 0.5°C of its set point. It also allows the RH of the internal air to be measured with satisfactory accuracy in the control band – say 85 to 95% RH – selected for unhindered growth combined with avoidance of the diseases associated with higher humidities.

A typical daily heating and ventilating profile for air temperature is shown in Fig. 4.10(b). The actual temperatures required will depend on the crop and the stage of its development but mostly night temperatures are set back with respect to day temperatures. Sudden changes of environment are avoided by automatic control of the set point to ramp up or down, as required, at the beginning and end of the day. The form and timing of the dawn/dusk ramps will be altered if thermal screens are employed in the house. These are plastic film sheets automatically pulled over the crop and round the side of the house at night, to reduce heat losses. Since they have a considerable effect on the heat and water balance within them, control settings have to be modified, too.

Relative humidity control further complicates this picture, rather as outside weather does. Although humidification and dehumidification equipment is available for RH control, its cost is usually too high for greenhouse production and the necessary regulation is achieved primarily by adjustment of the vents. However, limiting RH solely by increasing the ventilation may not be an option in external conditions of high wind, low temperature or high RH. Therefore, the controller may have to provide additional temperature lift – possibly up to 2 or 3°C – to reduce the RH in the house. In fact, to avoid excessively high RH the controller may have to override both the heating and ventilation algorithms and, in extreme circumstances, allow the heating and ventilation systems to operate simultaneously.

Yet another control factor can be introduced when carbon dioxide is added to the air in a greenhouse, to improve crop growth. Enrichment of the atmosphere to about 1000 vpm is economic at high solar radiation, when photosynthetic activity is enhanced. Although this process is often manually regulated, continuous monitoring and control systems are in use, and can ensure that wasteful or counterproductive injection is avoided. The system shown in Fig. 4.11(a) is based on the use of liquid bottled gas rather than the propane burners that are often used to generate CO_2. Air from the house or houses is drawn by a central pump through pipelines (via filters and condensate traps) at a steady flow rate and passed through a CO_2 detector such as that shown in Fig. 4.11(b) (see section 1.16). The sample lines can be as long as 200 m and each needs purging for about 1 min before a reading is taken. The meter itself can have a range of 0 to 3000 ppm $\pm 3\%$ of fro, and a time constant of 10 s. In this example a differential temperature sensor is used to monitor gas flow. Both elements are in contact with the gas but one is directly in the stream and the other is shielded from it. The sensor transmits an alarm signal to the computer if there is a blocked sample line. The computer can employ different set points during the day, according to the grower's programme, taking into account wind speed and radiation levels determined by weather sensors, as well as the requirements of the crop. Control can also be based on the differential level of CO_2 concentration inside and outside the house, rather than the absolute level. This is a safeguard against waste of injected CO_2 if the set point is near external levels and the gas becomes depleted in the house. Open ventilators are normally a cause of this waste, of course. In practice, there will always be a conflict between the needs of temperature control and CO_2 enrichment, since the latter is associated with higher radiation levels. Therefore another algorithm is required to limit the loss of the gas when the vents are open. One method is to reduce the set point as the leeward ventilators open, e.g. from 1000 vpm at 10% open to 300 vpm at 50% open.

Finally, there are times when the amount of moisture in the air has to be

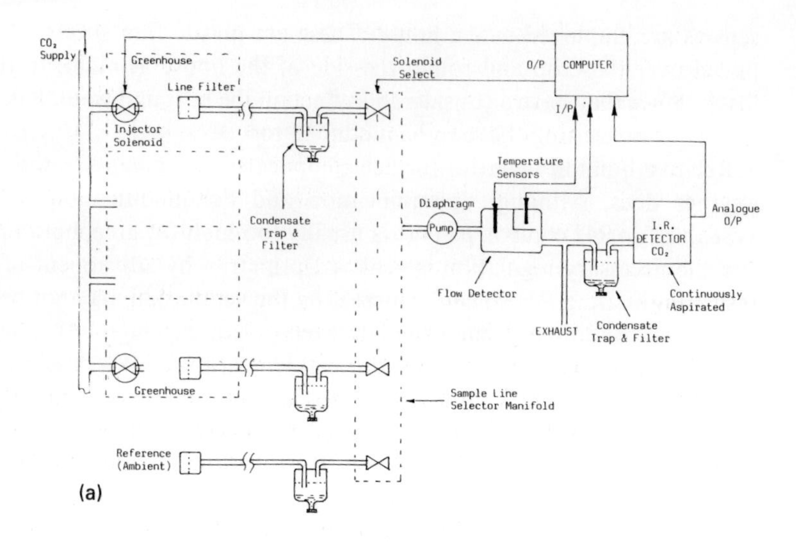

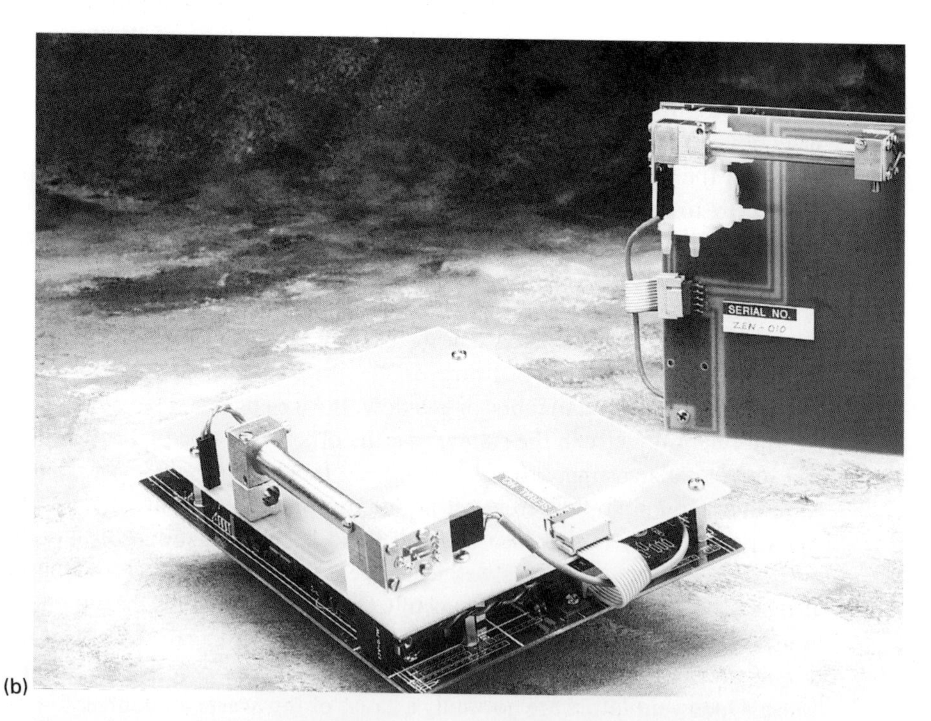

Figure 4.11 (a) CO_2 sampling and control system for multigreenhouse sites (Victor Automation Systems); (b) IRGA sensor and associated electronics (PP Systems).

enhanced. Humidification is necessary to encourage rooting of cuttings, for example. Closed loop control of the mist propagation process is based on the electronic *leaf* sensor which is placed among the cuttings. This unit is intended to simulate the drying out of the plant leaves and to switch on the mist spray as soon as a preset dry point has been reached. Physically, it is usually in the form of a short cylinder, mounted on a stem which can be pushed into the rooting medium, leaving the sensitive end face horizontal and exposed to the spray. This face has two flush electrodes (metal or carbon) embedded in ceramic, or in any electrically insulating material which can be wetted by the spray. While a water film remains on the face the measured electrical resistance of the leaf is low but the value rises quickly as soon as the surface begins to dry. At a preset resistance the leaf triggers a controller, which initiates the next burst of spray. The stand-alone controller in Fig. 4.12(a) provides a leaf sensitivity adjustment, which sets the resistance at which the spray is triggered. A second control allows the time of the spray burst to be varied from 0.5 to 30 s and a third sets the minimum time interval between bursts from 2 to 60 min. When the cutting is first inserted in the rooting medium the time interval is set towards the lower end of the span, and the cycle is largely, if not entirely, controlled by the output from the artificial leaf. If the resistance of the leaf drops below the switch-off point before the end of the time set on the BURST control, the leaf's signal overrides that control. Then, as the cutting matures, the setting of the INTERVAL control is increased progressively and/or the burst time reduced progressively, to wean the plants from the full mist regime.

A similar method is employed by fogging equipment, which produces a finer, more persistent cloud of droplets than mist nozzles. In this case the controller takes its input from a humidity sensor in the air. The action of the fogging nozzles is dependent on the provision of water under pressure, together with compressed air. No water can escape from them until air is applied and when the air is cut off a suck-back mechanism comes into play, to avoid water dripping on to the plants. Actuation of the nozzles is therefore effected by on–off control of a valve in the air line. A fogging burst is initiated when the output from the humidity sensor falls below the set point. One unit is capable of controlling several sets of nozzles in this way. As with the mist/wean unit, the time of burst can be preset within a given span (5 to 30 s, for example). It may also be possible to programme the controller to generate more than one burst/operation if required. The interval between bursts can be preset, too, up to about 1 h.

The root-zone environment

Here, available water and nutrients, together with the temperature of the rooting medium, are the quantities which may involve electronic monitoring and control. Rooting media include many different organic composts, inert organic materials and hydroponic solutions. Water may be applied on its own or as a nutrient solution (see Fig. 4.2). Taking straightforward irrigation first, there are many forms of open-loop sequence controllers which switch water on and off in different sections of a greenhouse, on a time basis. Incorporation of microprocessors

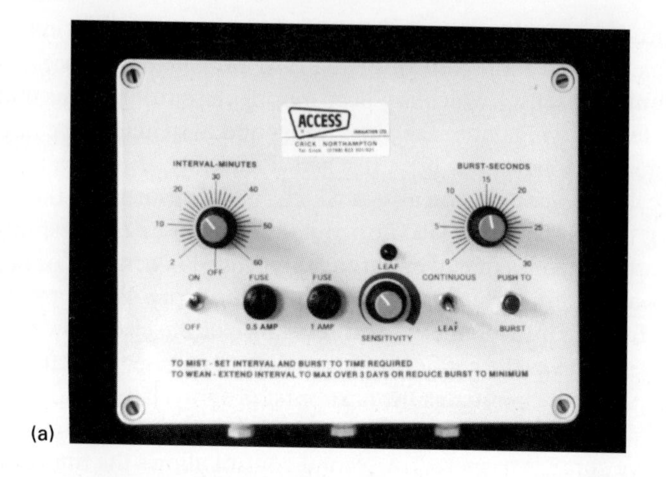

(a)

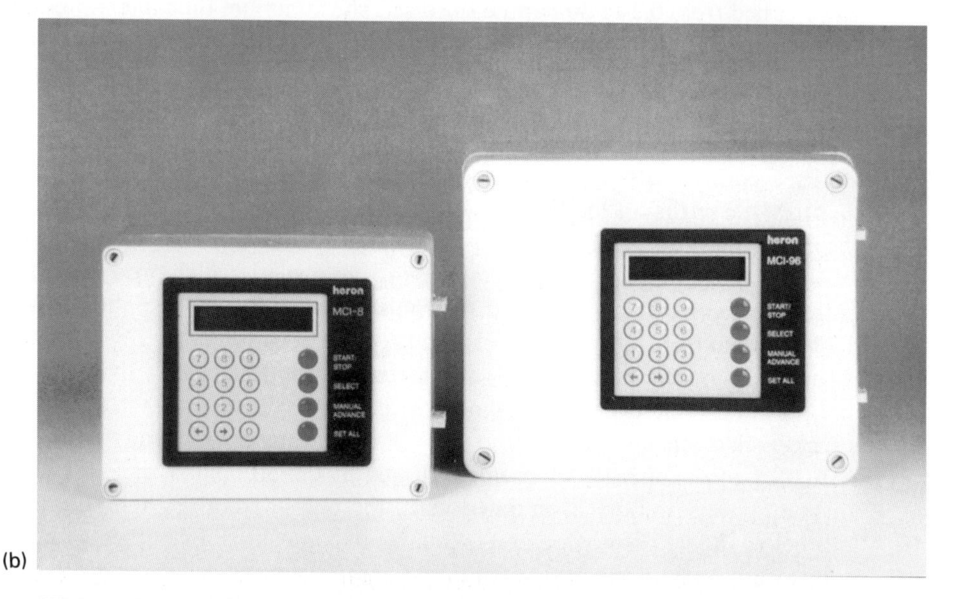

(b)

Figure 4.12 (a) Mist-wean controller; (b) irrigation controller (Access Irrigation).

can make them very versatile. The larger of the two units shown in Fig. 4.12(b) can operate 16 valves as standard, extendible to 96 by the addition of an expansion unit. Up to seven irrigation or misting programs can be supported, of which up to four can be run simultaneously. Program priorities can be assigned for times when one has to take precedence over another. Valve times can be set in the ranges 1 s to 60 min or 1 min to 60 h. Alternatively, by connecting a flow meter to the controller and entering the number of litres to be supplied by each valve, irrigation can be set up on a volume basis.

Up to 40 automatic starts can be assigned on a daily or weekly basis.

Alternatively, up to six automatic starts per day over a weekly period can be specified. Temporary increases or decreases in irrigation time across all programs can be effected by simply keying in the required percentage increase or decrease. If a time delay between valve operations is needed, that can be specified, too. The unit has two pump start outputs and a pump pressurisation period between 0 and 60 s can be set.

For cooling and misting, the controller has a continual cycling feature. This option allows the continual running of a program between specified start and stop times, with or without a delay between cycles. To support weaning applications, the controller allows for selected valves to be omitted from a program for a specified number of starts.

The unit can also be used with an EC/pH controller to deliver different EC levels for different irrigation programs. If it is coupled to a humidity or temperature sensor it can initiate continual cycling when the temperature rises or the humidity falls to a specified threshold. Cycling will continue until the condition has been rectified. When it is coupled to a light sensor, the controller can act as an integrator, initiating irrigation each time a total light energy threshold is exceeded (see section 3.2). Sensors are coupled to the controller via an RS485 interface.

The controller has a 5-year battery back up for its memory and a rapid response electronic circuit breaker on its output. If the latter is tripped by an overcurrent condition the controller displays the point in the program at which this took place. On the other hand, if no current is sensed in the line to a valve that has been activated the controller will ignore the condition or move to the next step in the irrigation sequence, according to the user's choice of setting. Remote start and stop facilities are provided. The unit can also be battery- or solar-powered, but not all the operational features are supported when it is powered in these ways.

More direct feedback of plant water requirements is available from infra-red thermometers and from evaporimeters, tensiometers and capacitance sensors, as in the outdoor environment (section 3.2). The last mentioned can be used in sand beds and capillary matting for watering pot plants, as well as in rock wool and other inert rooting media, employed for their ability to produce increased crop yield, quality and earliness at low energy cost, and for the absence of soil-borne diseases in these media. The plants grow in blocks of the medium, placed on the capillary matting. However, close control of nutrition is essential if these benefits are to be obtained, since all of the plants' requirements have to be supplied by the liquid feed which is taken up by their roots. This provides an application for nutrient injection equipment with pH monitoring, and large computer-based systems of this type are available for control of EC, pH and water temperature. The system shown in Fig. 4.13(a) combines controlled amounts of water, nutrients, acid and alkali in a mixing tank and feeds the plants by drip, circulation or ebb and flow systems. The computer can control the operation automatically in three ways – on a time schedule, in accordance with the measured level of solar radiation or on a closed-loop basis, by reference to the water loss in the rock

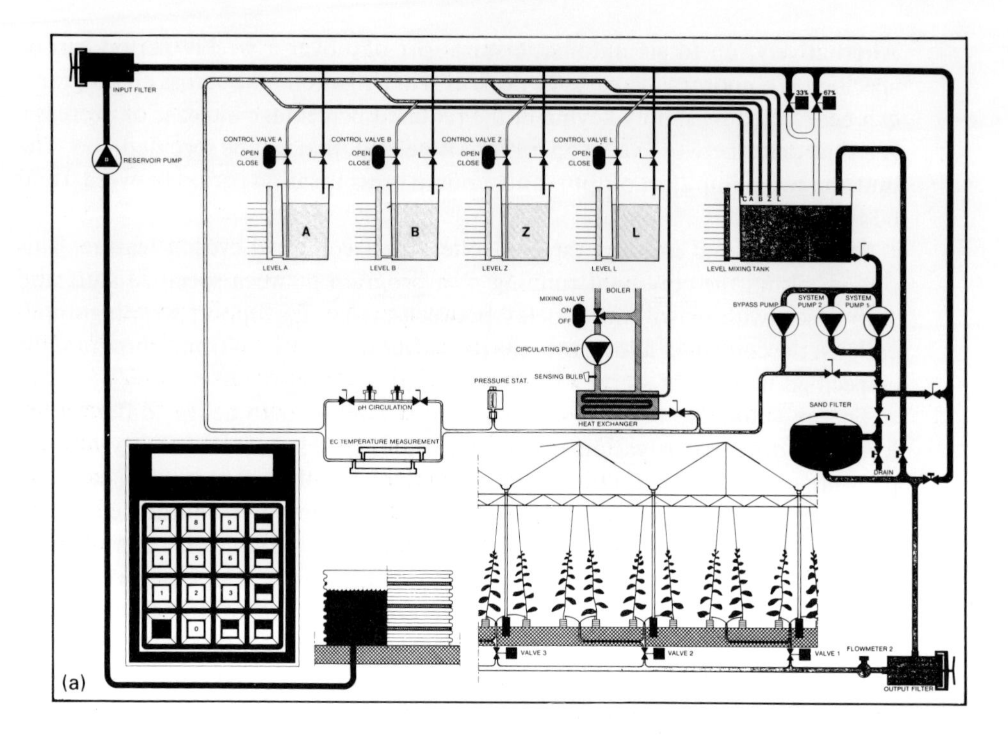

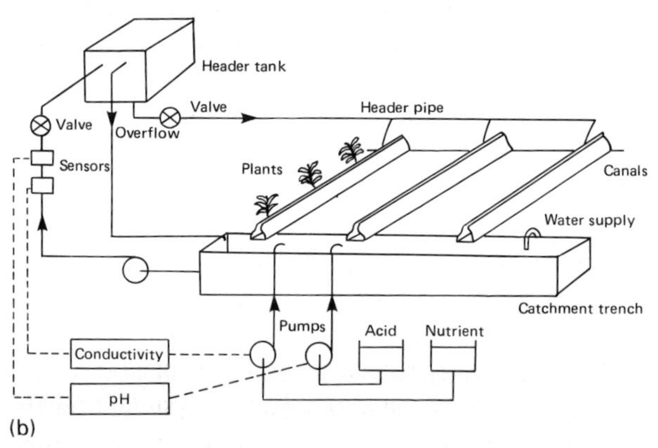

Figure 4.13 Liquid nutrients, greenhouse crops: (a) rock wool growth medium (Brinkman); (b) nutrient film culture.

wool. The EC and pH values for each block of plants in a large installation can be programmed separately. The salts content of the water supply is monitored to ensure close control of the feed properties. The computer also controls periodic back flooding of the sand filter, to prevent blockage. In view of the great risk to the plants if a malfunction should occur to the feed supply, high reliability has to be built into every component of the system. Nevertheless the computer is programmed to check all sensors and valves, as well as the conductivity, pH and temperature of the feed, before and during the feeding process. Alarm systems of several kinds can be activated if a fault is detected.

No solid rooting medium is required in the nutrient film technique of growing, shown diagrammatically in Fig. 4.13(b). Here the plants grow in parallel, gently sloping, plastic-lined ducts down which a water/nutrient mixture flows in a shallow layer, emerging at the lower end to fall into the catchment trench shown in the illustration. The tops of the channels are closed around the plants by a film plastic cover, which limits evaporation and keeps the channels clear. Dip-type or in-line conductivity and pH sensors are mounted in the pumped pipeline from the trench to the constant-head tank which supplies the inputs to the channels. The output from each is taken to a controller which adds concentrated nutrient salts or acid until the required balance is obtained. The set points vary from crop to crop and for a given crop the conductivity setting may vary substantially through the growing season. The nutrient liquid may be supplied only intermittently at early stages in crop growth since this can be advantageous culturally. The temperature of the nutrient feed is also monitored and controlled. One form of NFT controller employs a modular system, which incorporates adjustable conductivity, pH and temperature control units, high- and low-level alarm settings and timers. The timer module operates on the pumps which inject nutrient solution and acid into the catchment trench, to avoid overdosage. Optional modules allow control of dissolved oxygen level, pressure and flow conditions. A recorder can also be incorporated or the unit can be linked to a computer. Nutrient analysis is an obvious candidate for the application of the ion-selective sensors mentioned in sections 1.16 and 3.2. However, ion-selective electrodes do not have the robustness of conductivity electrodes, therefore any further development must depend on the introduction of suitably robust and specific ISFETs.

Integrated systems

The advantages of growing high-value crops under closely controlled conditions are well attested but something of the complexity of the necessary control systems can be appreciated from the foregoing parts of this section. The interactions between the external weather conditions, the plants' aerial and root requirements, and the heating, ventilation and irrigation systems employed call for a variety of interacting control algorithms and overriding commands. Other control functions may be required. Taking the system shown in Fig. 4.10(a) for example, when a single ventilation controller is used in a multicompartmented house with a separate vent drive in each ridge, the responses of the individual drives to the controller's commands are unlikely to be precisely the same. In consequence, over a period the vents in different ridges will become out of step. Therefore the controller must be programmed to close all the vents at intervals, to restore their initial synchronisation. Thermal screen control has been mentioned earlier but flowers and other crops may need shading during daytime, too, in order to avoid damage at high radiation levels. Similarly, regular blackout (photoperiod cover) is necessary to initiate flowering of some plants, such as chrysanthemums, at dates determined by the growers' market requirements. To additional time- or light-based controls such as these can be added further monitoring functions, among which measurement of heat input from pipeline systems is important as an indication of the fuel efficiency of a particular house operating under a particular environmental regime. This quantity requires the measurement of pipe temperature at the flow and return ends of the water circuit, together with measurement of its rate of flow. The product of these two quantities is a measure of the heat input. Since the temperature difference may be only a few degrees Celsius, platinum resistance thermometers are necessary to provide the required accuracy. Measurement of the hot water flow is not easy, either, but it is possible to use several devices, including the orifice plate (section 1.11), to provide an output which can be converted to an electrical signal. Then the measurement can be made electronically.

If the house is heated from an on-site oil-fired boiler, this too can be monitored and controlled electronically to maximise its efficiency. Boiler monitoring includes fuel flow, hot water temperature, temperature of the combustion air and temperature, oxygen content, carbon monoxide content and opacity of the flue gases. Maximal efficiency is attained when there is sufficient excess air entering the combustion chamber to avoid smoke emissions, as detected by the opacity (light transmission) of the exhaust gases. That condition results in an oxygen content of a few per cent in these gases, so providing an application for the high temperature zirconium oxide O_2 sensor that was introduced in section 1.16. In association with a catalytic combustion platinum resistance thermometer (PRT) the levels of hydrogen, carbon monoxide and other combustibles in filtered flue gases can be measured (to $\pm5\%$ fro), as well as oxygen, to detect incomplete combustion or defective burner equipment. These sensors can also be employed for stoichiometric control of the air/fuel ratio as the fuel supply varies to meet changing heat demands. However, the O_2 content of fuel gases is closely related to their CO content, and the latter is a more sensitive indicator of the optimal ratio. Therefore the use of an infra-red gas analyser for carbon monoxide is sometimes preferred.

Another essential feature of integrated systems is the provision of alarm systems through local audible alarms and telephone call-out facilities.

Overall, the advantages of integrated systems have been realised since the 1980s and many well established commercial forms now exist. Figure 4.14 is a composite of their different forms. Together they have the SCADA structure (Fig. 2.5) mentioned earlier in this section. The central climate controller/PC, with an input from a local weather station and interchanges with sensors and actuators in the house, is probably the most common system. In a distributed system the PC communicates with local climate control units, located either in separate compartments of a house or in separate houses. These are linked by a data bus. RS422 and Ethernet (section 2.2) have been employed for this purpose, since bus lengths can approach 1500 m. The full system has a PC in which optimal control models can be developed and applied as supervisory control. This PC is shown in Fig.

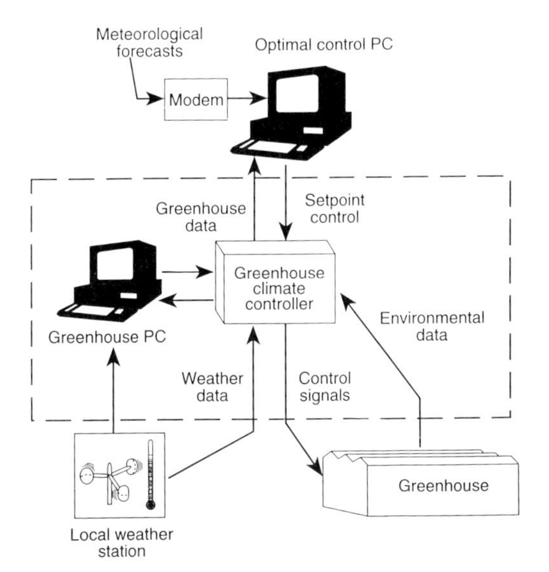

Figure 4.14 Integrated greenhouse control system (Silsoe Research Institute).

4.14 with a modem link to a meteorological forecasting station, because with that input optimal environmental control settings can be calculated to take account of predicted temperatures and wind speeds, before downloading them to the local climate controller. In fact, the PC's link to the climate controller can be a modem and telephone line. The supervisory PC can also maintain expert systems of different kinds, for diagnosis of growth problems, as well as for economic models, updated from market intelligence.

Microplants

Large-scale micropropagation of plantlets of a variety of species – from ornamentals to forest trees – has given rise to developments of image processing which have involved many types of greenhouse plant. These have been devised to provide machine vision for the mechanisms which take cuttings from the mother plants, and for the harvesting mechanisms which transplant the viable microplants. Figure 4.15 illustrates the first operation. It shows a binary image of a chrysanthemum plant, i.e. all pixels outside its borders have been set to white and all those inside to grey. After this thresholding process the computer algorithm has searched for and located the nodes, as can be seen in the figure. From that information it can guide the robot arm which cuts and removes the segments of the plant and inserts them in the culture medium.

At the harvesting stage the system is presented with groups of plantlets of similar structure to the mother plant, rooted in their container. Another algorithm then outlines their stems, for guidance of a lifting and transplanting mechanism.

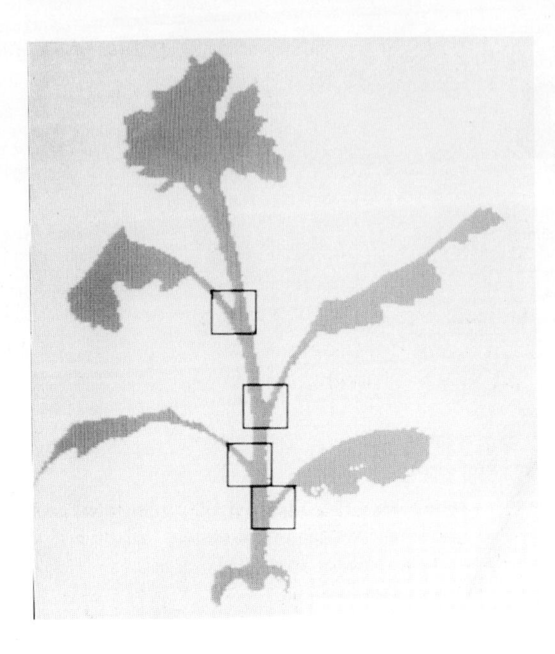

Figure 4.15 Computer imaging in microplant production. Identification of nodes for cutting (Silsoe Research Institute).

Mushrooms

Mushroom production involves four stages, namely preparing the compost, pasteurising it, mixing it with spawn for *spawn running* over 10 to 14 days and, finally, capping and putting it into the growing room. The last three stages require the material to be held at specific temperatures and the last two require monitoring of ventilation, RH and CO_2. Portable hygrometers and anemometers are convenient monitors, while environmental controllers of the type familiar in crop stores (Chapter 5) have been adapted to the mushroom growers' requirements. Ventilation fans, heaters and humidifiers are controlled to achieve set levels of air temperature, airflow and RH during the production cycle, and for heat sterilisation of the compost after harvesting. Carbon dioxide is also monitored (by the ubiquitous IRGA), over the range 1000–2000 ppm.

This is an appropriate point to note that a new class of sensor has been added to the ISFETs of section 1.16 as a means to create *electronic noses*. It is in the form of an array of fast-reacting, conducting polymers, with low power requirements and the ability to respond to many polar, volatile chemicals. A 20-element array has been developed to locate underground truffles (a task hitherto performed by trained dogs and by pigs). The instrument is a hand-held device, akin to a metal detector, which sucks air from near ground level over the array. It incorporates pattern recognition software to analyse the sensor outputs and to seek for clusters that resemble those produced by truffle odours. The array also responds to air humidity since it is water sensitive, but the cluster pattern of water differs from that of the truffle, so the distinction can be made. Tests over 2 years showed that the sensors made reliable decisions, although only down to 100 mm depth. These sensors are capable of sensing odorants at levels below 1 ppm and have evident relevance to livestock operations (Chapter 6).

4.5 Further information

Selected papers on the more advanced techniques described in this chapter are given below. Most provide leads to other significant papers, for those who wish to go more deeply into these subjects. For further information on colour and colour standards, specialist optical texts are strongly recommended; although some material is accessible on the WorldWideWeb.

Section 4.2

Marchant, J.A. (1996) Tracking of row structure in three crops using image analysis. *Computers and Electronics in Agriculture*, **15**, 161–79 (autonomous vehicle).

Noyman, Y. & Schmulevich, I. (1996) Ground surface sensing through plant foliage using an FM-CW radar. *Computers and Electronics in Agriculture*, **15**, 181–93.

Palmer, J., Kitchenman, A.W., Milner, J.B., Moore, A.B. & Owen, B.M. (1973) Development of a field separator of potatoes from stones and clods by means of X-radiation. *Journal of Agricultural Engineering Research*, **18**, 293–300.

Section 4.3

Plá, F., Juste, F. & Ferri, F. (1993) Feature extraction of spherical objects in image analysis: an application to robotic citrus harvesting. *Computers and Electronics in Agriculture*, **8**, 57–72.

Plá, F., Juste, F. Ferri, F. & Vicens, M. (1993) Colour segmentation based on a light reflection model to locate citrus fruits for robotic harvesting. Special issue: computer vision (eds J.A. Marchant & F.E. Sistler). *Computers and Electronics in Agriculture*, **9**, 53–70.

Slaughter, D.C. & Harrell, R.C. (1989) Discriminating fruit for robotic harvest using color in natural outdoor scenes. *Transactions of the American Society of Agricultural Engineers*, **32**(2), 757–63.

Section 4.4

Bakker, J.C., Van den Bos, L., Arendzen, J.A. & Spaans, L. (1988) A distributed system for glasshouse climate control, data acquisition and analysis. *Computers and Electronics in Agriculture*, **3**, 1–9.

Chalabi, Z.S., Bailey, B.J. & Wilkinson, D.J. (1996) A real time optimal control algorithm for greenhouse heating. *Computers and Electronics in Agriculture*, **15**, 1–13.

Davis, P.F. (1991) Orientation-independent recognition of chrysanthemum nodes by an artificial neural network. *Computers and Electronics in Agriculture*, **5**, 305–14.

McFarlane, N.J.B. (1991) A computer-vision algorithm for automatic guidance of microplant harvesting. *Computers and Electronics in Agriculture*, **6**, 95–106.

Persaud, K.C. & Talou, T. (1996) Hunting the black truffle. *Land Technology*, **3.1**, 4–5.

Seginer, I., Boulard, T. & Bailey, B.J. (1994) Neural network models of the greenhouse climate. *Journal of Agricultural Engineering Research*, **59**, 203–16.

Chapter 5
Post-Harvest Treatment of Crops

5.1 Introduction

This chapter brings together crop weighing, handling, drying and storage, followed by quality determination and grading. These are all areas crucial to the value of crops in quality-driven markets, where food processors and supermarket chains have a powerful influence. In consequence, the need for automatic data logging and control of the progress of crops from the field to the market has assumed increasing importance.

5.2 Crop weighing and handling

Crop weighing

Chapter 3 introduced in-field weighing of harvested cereals and vegetable crops by direct and indirect means. In this section the subject is static platforms for weighing trailer and truckloads of produce (and other materials) on to or off the farm. Even when a weighing trailer such as that shown in Fig. 3.20 is employed, its load readings should be checked against those of a fixed weighbridge or weigh platform from time to time. The emphasis on *fixed* is not intended to discount the value of portable, battery or mains-powered weigh pads. These are robust devices, with a suitably low profile (e.g. 100 mm overall height), which can be placed on any level, firm surface where they can support the weight of a single wheel. In pairs, spaced at the right distance, they form an axle weigher. Some can be used dynamically (i.e. the vehicle or trailer can be driven over them slowly), but they are normally used for static weighings. Each pad contains a low-profile load cell with a capacity in the 5–15 t range. Where two or four are used as axle/vehicle weighers the meter can totalise the measured weights. The weighing system may be capable of 10 kg resolution but the accuracy attainable depends on the setting up of the pads and the firmness of the supporting surface, as well as the size of the load, *inter alia*. The separation of the pads has to be adjusted to each different axle width, of course. Therefore, for general farm use permanently installed, single- and multi-axle weigh platforms are more in favour, since the siting and levelling procedure is done once and for all.

With these mains-powered units routine weighing of trucks and trailers becomes practicable, to assist store management, to measure crop yields and to check individual axle weights before laden trucks move on to public highways. Weighing of a tractor and trailer, or a truck, normally takes less than a minute. The weighers may be designed for static use only or for dynamic weighing as well. Their width is normally 3 m and their length depends on the number of axles that they are capable of weighing simultaneously. Their capacities are generally in the range 10 to 30 t, with overload capability of 30% or more. The platforms themselves can be decks of steel, or steel frames filled with concrete (normally poured on site). Some steel platforms are transportable; some steel and concrete ones are fixed but surface-mounted, with lead-in and lead-off ramps. However, most are mounted in specially prepared pits.

Their performance depends initially on the accuracy and reliability of the summated output of the four corner load cells that are normally employed to support the platform. These are generally the strain-gauged transducers introduced in section 1.14, in the form shown in Fig. 1.7, since it is essential to minimise the effects of off-centre application of their loads and any side loads. The cells must be temperature compensated and sealed against ingress of moisture and dirt, since they are working in exposed conditions. Their connections to the associated meters/printer units (usually housed in a neighbouring office or shelter) must be fully environment proofed, too. Next, the site must be level because the load cells will only measure the loading along their sensitive axis. If the platform is not horizontal and their loading axis therefore not vertical they will measure less than a load's true value. Furthermore, if a truck or a tractor and trailer are weighed in successive stages (e.g., tractor, then trailer, or axle by axle) then there must be a continuing length of smooth level surface at the front and rear of the platform. This is necessary to ensure that all of the surfaces supporting the axles are in the same plane. If this is not achieved it will affect the weight distribution on the separate axles and will introduce error into the summation of axle weights. Therefore site preparation requires careful laying and levelling of considerable amounts of heavy duty concrete. Equal care is necessary in the preparation of the pit, since the platform should be flush with the surrounding concrete and the load cells must share the load as equally as possible, to eliminate corner errors. Proper drainage of the pit is essential (see Fig. 5.1(a)) and the drainage system must be designed so that it does not silt up. Even so, maintenance (or maintenance contracts) should include occasional lifting of the platform assembly for inspection of the base of the pit.

The quoted accuracies for a properly set up and maintained installation are usually in the range $\pm 0.5\%$ to $\pm 1\%$ of fro. That quantity should not be confused with the rated capacities of the load cells. The dead weight of the platform introduces a tare weight and allowance must be made for possible overloads, therefore the working range of the cells is considerably shortened. In addition, it must not be forgotten that the accuracy will be lower at less than full range load.

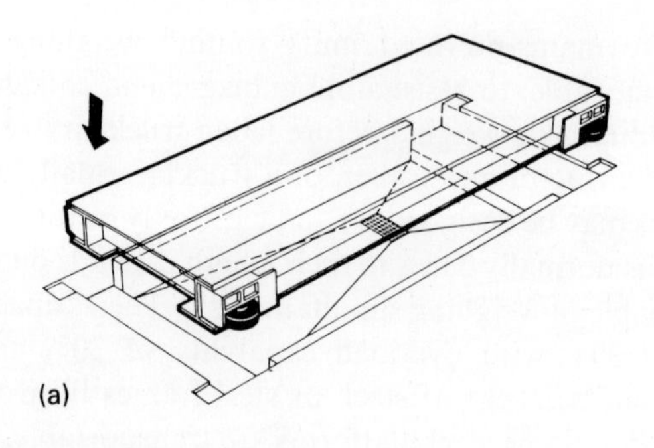

Figure 5.1 (a) Weighbridge pit (Weighwrite); (b) calibration weighing (Griffith Elder).

For example, if the quoted accuracy is ±0.5% at 10 t the actual error is ±50 kg. That is equivalent to ±2% at 2.5 t load.

Modified traffic lights are available with some commercial axle weighers, to help the driver to locate each axle over the platform in turn. Similarly, they are used to warn drivers when they are exceeding the speed limit over dynamic weighers. The meters need time to process the load cells' signals and that is minimal if there are close-coupled axles on the trailer or truck. Speed limits are therefore of the order 3 to 4 km/h.

Both static and dynamic weighing require a knowledge of the vehicle and trailer weights in order to derive the weight of the crop. Figure 5.1(b) shows the determination of these values in progress. This calls for careful recording if several tractors and trailers, or trucks, share the same weighplatform throughout the day. That is no problem for large-scale stores or processing plants equipped with automatic identification of vehicles, data bases with details of their tare weights, owners etc. However, simpler version of these systems are available to individual farmers, enabling the driver to transmit details of the truck or tractor/trailer combination to the weight recorder without leaving the cab. This can be done by an infra-red transmitter or by a transponder system such as that used with animals (see Chapter 6).

Indoor conveyor systems also include weighers of several kinds. Some are similar in principle to the volume measuring devices used to measure yield on combine harvesters (section 3.3). Continuous belt weighers have provided another application for load cells. Others are in the form of batch weighers, often required to conform to weights and measures standards for trading purposes. Many of these are essentially mechanical balance systems, coupled to electrical switches and actuators, but check weighing has provided an application for the LVDT (section 1.9). This is the sensing element in a flexure-pivot weighcell shown diagrammatically in Fig. 5.2.

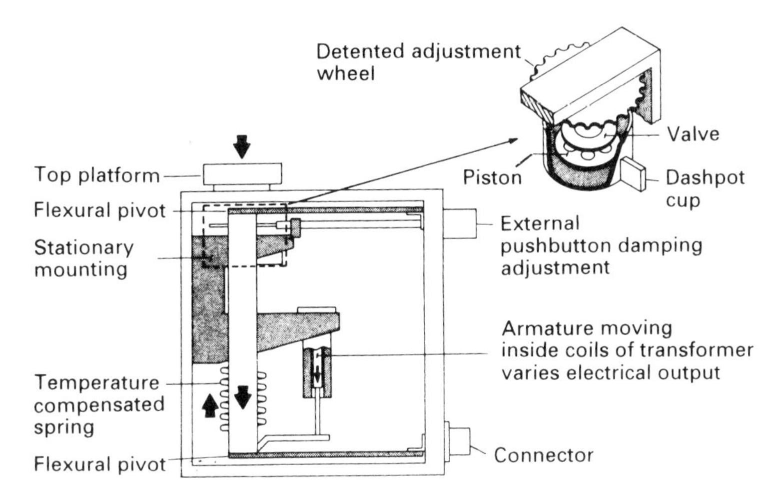

Figure 5.2 LVDT weighing unit (Hunting Electrocontrols).

A load on the weighplatform applies a downward thrust on a frame supported on frictionless flexural pivots at its top and bottom. Movement of the frame is resisted by the temperature-compensated spring until, as in all spring balances, the change in length of the spring (in this case, compression) creates a force which balances the weight of the load. The resulting downward movement of the frame is measured by a LVDT, with the coil assembly fixed to the stationary mounting, and the core fixed to a plate which is attached to the frame. The small lateral movement of the frame which accompanies its downward movement – due to its method of suspension – is of no significance, because its maximum depression is 1.5 mm. However, this limited travel calls for a very sensitive linear displacement transducer. For dynamic weighing the combination of a compression spring and a mass (the load, together with the mass of the platform and frame) requires a degree of damping to minimise oscillation and overshoot under transient and shock loads (Figs 1.3(b) and (c)). The damping employed in this cell is the traditional piston moving in an oil dashpot (which means that the cell must not be tilted). This can be adjusted to critical damping in particular circumstances by operation of the external push button, which controls the damping valve via the detent wheel.

Another adjustment, not shown in the diagram, enables the user to preload the system (i.e. set a fixed tare) without unsealing the cell. The tare can be several times the cell's span. For example, the smallest cell of the type shown has a total capacity of 885 g and a live span of 125 g. This means that a dead weight load of up to about 750 g can be offset, leaving the cell to work normally over its 125 g span. Spans up to 16 kg are available and all have an overload limit of 225 kg which is set by mechanical stops. Their linearity and repeatability are $\pm 0.1\%$ and $\pm 0.02\%$ of span, respectively, and temperature effects between 6 and 50°C are less than $\pm 0.01\%$ of span. They can also be used with off-centre loading, with a maximum error of –0.25% in spans up to 4 kg, and –0.75% in spans up to 16 kg. Their response time is 30 to 200 ms, depending on span, tare load, damping and loading method. Higher spans give faster response and higher tares give slower response. The manufacturers provide an approximate equation for calculating the time required by the cell to approach its final output within 0.5%, and without overshoot, when a load is applied. Thus:

$$\text{time (ms)} = 70\sqrt{(\text{tare} + 540)}\ /\text{span}$$

when both tare and span are in grams. The performance of these cells is reflected in their cost, of course, but no sensor for check weighing can be inexpensive.

Crop handling

Materials handling is a major activity in all areas of agriculture and horticulture, embracing bulk handling, packaging and, in the horticultural context, movement of plant containers in many shapes and sizes. Several common sensors contribute to the control of materials flow in a variety of applications, some of which are introduced in Chapter 6, because they relate to livestock operations. In the arable sector, proximity sensors are employed to control the height of elevator spouts in order to limit the fall of damageable produce being discharged into a bulk container. The ultrasonic height sensor mentioned in Chapters 3 and 4 has been used in this way with potatoes, to keep the spout just clear of the top of the pile. Similar sensors are used in potato bagging equipment, where capacitance sensors have also been used.

Damage suffered by vegetable and fruit crops during handling operations is of great concern to producers, in view of its effects on quality assessment, therefore identification of sources of damage is an important matter. For that reason several types of artificial fruit and vegetables have been built to undertake the detective work. These pass through the materials handling system, recording pressures and impacts, and storing a time history of the loadings to which they are subjected. Figure 5.3(a) gives an outline of a recent example of the species. The principal requirement of these devices is that they should be of approximately the size, shape and elasticity of real produce. However, the compromise design is usually a deformable sphere with essentially equal sensitivity to pressures and impacts from any direction. The outer sphere in Fig. 5.3(a) is a 62 mm diameter rubber ball with a 4 mm skin, containing a spherical instrumentation module, centred by conical springs. Any deformation of the outer sphere is communicated to the inner sphere as an increase in the pressure of the intermediate silicone oil. The whole of the electronic circuit is encapsulated in epoxy resin but the channel shown in the diagram allows a pressure sensor to respond to the changes in oil pressure. Temperature compensation is provided by an adjacent temperature sensor.

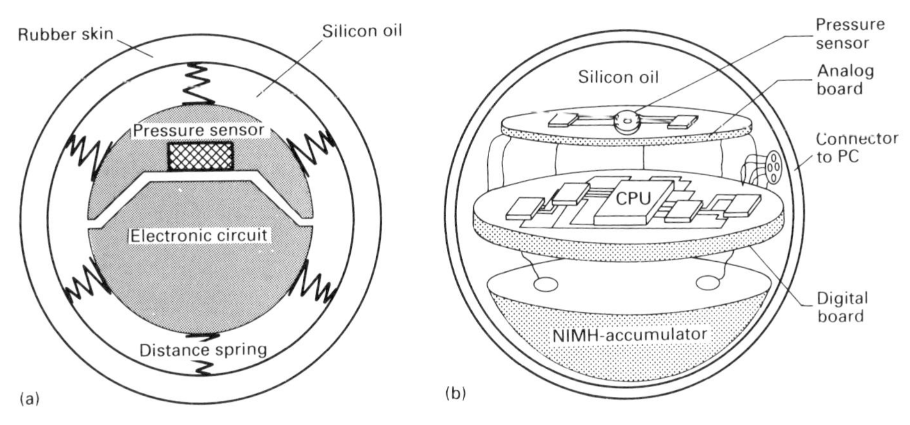

Figure 5.3 Artificial vegetable: (a) general construction; (b) inner sphere (Institute of Agricultural Engineering, Bornim).

The encapsulated circuit (Fig. 5.3(b)) contains a rechargeable nickel–metal–hydride battery. This can be recharged via the 5-pin connector on the outer shell, which also provides the means for an external PC to program the CPU and to collect the stored data via the RS232 interface. The sampling frequency can be adjusted to any value between 4 Hz and 10 kHz and the computer can be programmed to store pressure peaks above a chosen threshold level. At the start of a run the CPU's real-time clock is synchronised with that of the PC. Then an observer can put time markers into the PC at times of particular interest as the sphere progresses through the handling system. When the sphere's collected data are downloaded to the PC the markers appear on the display, at the appropriate place in the time series of peaks collected by the sphere.

Cereals do not present the same damage problem but large grain handling and drying installations need comprehensive supervisory control, to ensure that conveyors, elevators, high-temperature driers, etc., operate safely, at the right time and in the correct sequence. In the event of a problem, such as a fuel supply or burner fault, overheating of the grain, or a flow blockage, the installation must be shut down as quickly as possible, but, again, in the right sequence. This is an ideal application for the programmable logic controllers (PLCs) introduced in section 2.3, coupled with level, load, motion and temperature sensors of conventional kinds. Figure 5.4 shows a control unit for a grain handling facility, with a mimic diagram of the installation at the top, the main switching unit in the centre and a PLC with two extension units at the bottom (cf Fig. 2.7). The mimic diagram has indicator lights to show which elements are active at a given time. In this experimental facility an underfloor conveyor with three access gratings feeds grain to the base of the central vertical elevator and from there into any of four holding bins. From two it can be elevated to an overhead conveyor, from which it can be directed to the cleaner, a grain conditioner or the high-temperature drier. Table 5.1 lists the PLC's facilities and the inputs and outputs employed. Programs were set up on the program loader on the left of the PLC system, using its array of 42 buttons. The main program required about 300 steps, out of the maximum capacity of 1024. Monitoring of air flows, temperatures, weights, etc., is the function of a separate computer in this installation.

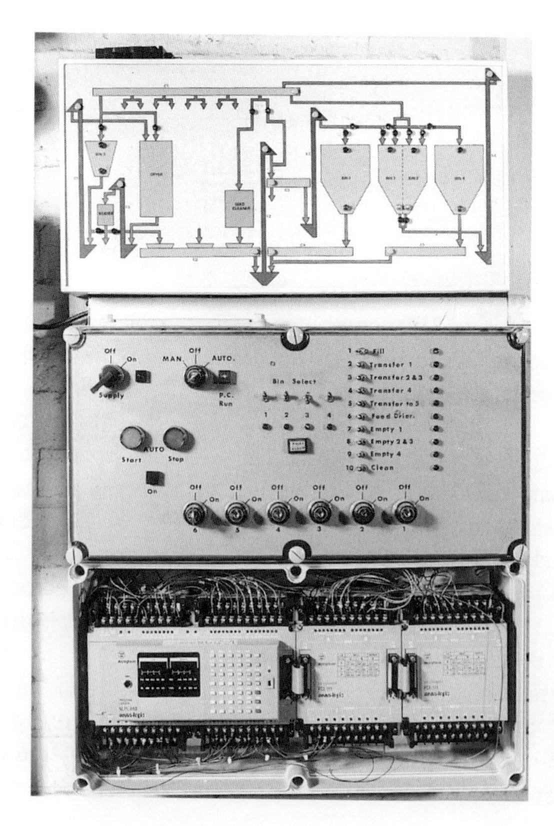

Figure 5.4 Programmable logic control system for a grain drying and handling installation (Silsoe Research Institute).

Table 5.1 Functions of the programmable logic controller shown in Fig. 5.4 (Silsoe Research Institute).

Inputs (up to 56)	
• Bin selector switches	4
• Function switches	10
• Diverter sensors	14
• Bin full sensors	5
• Bin empty sensors	5
• Motor overload sensors	8
Outputs (up to 32)	
• Motor contactors	8
• Solenoid valves	2
• Alarm	1
• Diverter relays	7
Timers 0 to 12.7 s	
Counters 1 to 127	

5.3 Crop drying and storage

Grain moisture determination

Safe and efficient harvesting, drying and storage of grain are dependent on means to measure its water content reliably between field levels and safe storage levels. With regard to the latter, Table 5.2 summarises the relationship between the moisture content (wet basis), storage temperatures and safe storage life for the two main cereal crops. [*Note* Percentage moisture content, wet basis (or mcwb) is defined as

$$\frac{m}{m + M} \times 100\%$$

where m is the mass of water in a combined mass of water and grain (M). It is quoted to one decimal place only.]

Standards

Water in crop material can exist in several states. It can be present as free superficial moisture, as absorbed water in the pores of material, as a tightly retained (or bound) adsorbed layer on the crop tissue and as chemically bound water in the tissue. The ease with which the water can be extracted from the material decreases throughout the above progression. Therefore there is a question of definition: how many of these states of the water should be regarded as a part of the material's moisture content? In fact, chemically bound water is usually excluded, but the boundary is ill defined. There is also a problem of measurement. Any method that seeks to measure the moisture *in situ* is faced with the inevitable differences in response of water in different states of binding, in

Table 5.2 Estimated maximum safe storage life (weeks) (ADAS).

Barley

Storage temp (°C)	Moisture content % (wet basis) 11	12	13	14	15	16	17	19	23
20	110	80	50	32	19	10	5	2	0.5
10	600	400	260	160	90	50	21	8.5	2

Wheat

Storage temp (°C)	Moisture content % (wet basis) 12	13	13.5	14.5	15.5	16.5	17.5	19.5
20	56	40	28	19	13	7	3.5	1.5
10	200	140	95	60	38	20	11	4.5

addition to any effects due to other constituents of the material. Therefore, the most reliable methods for moisture determination have always been seen as the extraction of water from the material in an oven. Even then, there are difficulties to be overcome. Overcooking of the material can remove volatile components other than water. Therefore, numerous national and international committees, backed by equally numerous research and testing laboratories, have laboured to establish acceptable national and international standards of oven drying, for particular materials and particular purposes. The crops of most common concern are cereals and oil seeds, hence it would be advantageous to have common methods of determination for them. Unfortunately, there is no universally agreed method yet, although ISO has established two standards which are widely used. The differences between these standards relate to the drying temperature, drying time and use of ground or whole samples of the seed. These differences yield different results, therefore direct comparisons of moisture content determination are not possible if two different methods have been employed.

For example, ISO R712-1979, the international standard for grain trading, requires a ground, 5 g sample, an oven temperature of 130 to 133°C and a drying time of 2 h. In contrast, the ASAE standards for wheat and barley require whole 15 g samples, dried at 130°C for 19 and 20 h respectively. However, there are some common requirements which can be summarised here. Standard oven determination requires replicate measurements on subsamples of seed representative of the sample under test (a standard procedure is laid down). These are placed in specified containers with close fitting lids, which have been thoroughly dried in a desiccator and then carefully weighed, to the nearest mg. If the subexamples have to be ground, this operation must be carried out precisely to the requirements of the standard. When the requisite amount of the material has been put in each container and spread evenly within it, the lid is quickly placed on the container and the whole weighed again. The samples, with the lids removed, are then placed on shelves in a thermostatically controlled oven, with good natural ventilation and temperature of specified uniformity in the region of the samples. After drying for the prescribed period, the lids are replaced on the containers and the samples placed in the desiccator until they reach laboratory temperature again. They are then reweighed. The loss of weight is equated to the amount of water removed by drying, in grams. Then the mcwb is derived. All crop moisture meters must be referred to one or other of these standards if their indications are to be meaningful, and the standards employed should be quoted in the manufacturer's literature, since there can be differences of up to 2% mcwb between them. This entails extensive comparisons between their readings and oven determinations, for each crop with which they can be used, and over their temperature range. Furthermore if they are to be used for trading purposes they must contend with the specifications laid down by the Organisation Internationale de Métrologie Légale (OIML), which is the body governing international trading standards. These specifications appeared as a Recommendation in October 1984 and have since been ratified internationally. For wheat, oilseed and some other seeds the specification limits error to ±0.3% mcwb, at and below 10% mcwb, and to ±3% of the meter's indication above that level (e.g. a maximum error of ±0.6% mcwb at 20% mcwb).

Moisture meters

Methods of measuring grain moisture content are manifold. From the standard oven drying methods, through RH measurement, distillation in heated oil, infrared drying, neutron moderation and nuclear magnetic resonance, to measurement of the grain's electrical properties by varied means. However, in general practice, two well known types of grain moisture meter now predominate, by virtue of their ability to meet the user's requirements with speed, simplicity in operation, portability and reliability. One type measures the electrical resistance of a ground, compressed sample, while the other measures the capacitance of a sample of whole grain at a MHz frequency. Both facilitate the systematic multiple sampling required during bulk drying and storage of grain (Table 5.3).

The resistance method has a long history, starting with the development of the Marconi moisture meter in the mid 1940s. That introduced the routine of milling a clean sample, placing it in a cylindrical cell with two annular electrodes in its base, compressing it with a close-fitting piston via a screw clamp, then reading an analogue meter. Separate scales were used for different types of grain, covering the approximate range 10–25% mcwb. Corrections to the readings had to be made if the ambient temperature was not the same as the calibration temperature of the meter. Data were supplied for this correction, which is approximately 0.1% mcwb/°C. The meter's operating temperature range was 0–40°C.

The annular electrodes formed part of a d.c. bridge circuit. The resistance between them depended essentially on the moisture content of the grain close to

Table 5.3 Bulk drying and storage of grain: sampling.

Batch size	Minimum no. of samples
300	30
250	25
200	20
150	15
100	10
10	5

them, therefore the amount of grain in the cell was not critical to the measurement, provided that the sample was well mixed in the grinding process and well compacted by the screw clamp. The main problem with the measurement at that time was the variation of resistance with moisture content, which is vast and highly non-linear (see Fig. 5.5(a)). Above 30% mcwb it could be only a few ohms but at the dry extreme it could rise to 100 MΩ. It required the use of a special type of thermionic valve to cope with the latter resistance.

However, semiconductor electronics has brought about many changes to the capabilities (and shape) of this type of meter. Figure 5.6(a) shows a unit of the mid 1990s, with an integral grinder and compression cell which locks over the spigots surrounding the central electrode assembly. The arm on the top of the grinder/compressor has a central slider which is set to one position for grinding and another for compression. In the latter mode a clutch slips and clicks when sufficient compression has been achieved. The whole assembly can be quickly disassembled for cleaning. In use a push button brings up the reading of the sample moisture content in the upper LCD display, automatically corrected for variation from the 20°C (European standard) calibration temperature. The lower LCD indicates the selected seed calibration, together with the temperature of the measurement. The instrument has ten standard calibrations programmed in its memory, including maize, oil seeds, peas and beans as well as cereals. The user can allocate four frequently used calibrations to *short cut* keys. The battery-backed, non-volatile memory retains the last calibration selected until the meter is reused. In addition, it has an averaging facility for multiple measurements and an RS232 output to a serial printer or PC. New or up-to-date calibrations can be added by the dealer and user, as well as the manufacturer, up to the memory's maximum capacity of 50. Finally, the meter has a connector socket which enables it to monitor temperatures in stored crops via a thermistor probe, extendable to 4 m.

A similar, hand-held version employs a single rotary knob which is used to balance the meter electrically. The balance point is indicated by an LED adjacent to the knob. The angular position of the knob's pointer is then converted to a moisture content reading by reference to a scale ring surrounding it. A readily

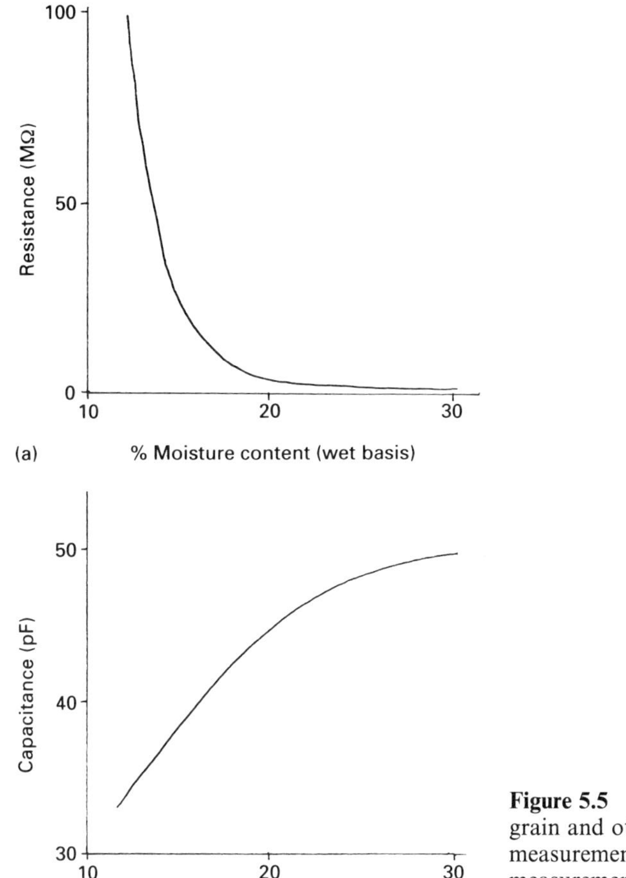

Figure 5.5 Moisture measurement in grain and other seeds: (a) resistance measurement; (b) capacitance measurement (Silsoe College, Cranfield University).

interchangeable set of rings is available for different crops, each indicating the oven standard on which it is based.

The capacitance meter depends on the dielectric properties of water – exploiting its high relative permeability, as already stated in section 3.2 (under Electronic instruments). It is necessary to perform this measurement at a frequency in the MHz range because with an unground sample the distribution of moisture within the grain sample affects the measurement. Any degree of surface wetness or dryness relative to the interior of the grain affects the conductance element of its complex impedance to an extent that is more pronounced at lower frequencies. A second consequence of the use of unground grain is that the packing (bulk) density of the grain can vary, therefore the cell must be filled as uniformly as possible, with a known mass of the grain. As with the resistance method, the calibration varies with the type of grain – and even for a particular variety it can change from year to year. A similar temperature coefficient also

(a)

(b)

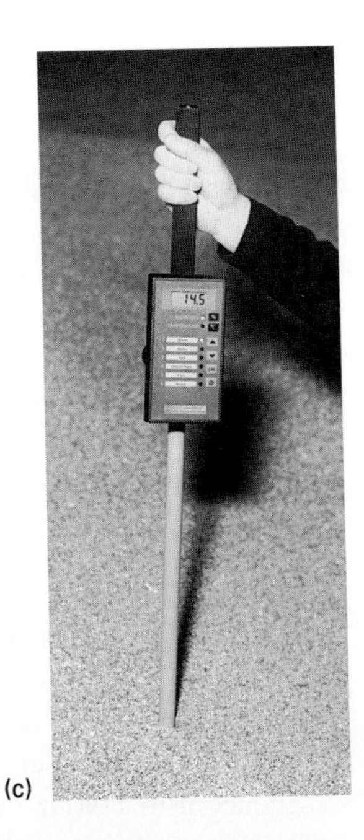

(c)

Figure 5.6 Moisture meters: (a) resistance measurement (Protimeter); (b) capacitance measurement (Sinar); (c) moisture and temperature probe (Sinar).

applies. The capacitance vs mcwb relationship is equally non-linear but in the reverse direction (Fig. 5.5(b)). However, the result is the same – there is diminishing sensitivity to changes in moisture content beyond 25% mcwb, while it becomes increasingly difficult to make meaningful measurements towards 0% mcwb because of the rapid changes of sensitivity in the lower region.

One of the many forms of portable, battery-powered capacitance meters is shown in Fig. 5.6(b). It has an annular cell, with a capacity of 290 ml. This provides for a representative sample of the seed, although users wishing to test a hand-rubbed sample in the field may prefer a meter with a smaller cell. Like all capacitance meters designed for close comparison with oven methods it has a range of necessary features. One is a cell-weighing system for measurement of the sample's mass. In this case the cell and its load are mounted on a spring element and the mass–spring system is set in vibration electromagnetically. The frequency of the oscillation is measured digitally and that is a measure of the mass. This method has the considerable merit that the meter is not required to be horizontal when the measurement is made. The meter can measure and adjust for mass between 20 and 240 g in this way.

The cell still has to be filled as uniformly as possible, therefore the load is dumped from a hopper which fits over the cell. The grain is retained in the hopper by a spring-loaded slide with an off-centre aperture, actuated from the side. When the slider is pressed the grain is released rapidly into the cell's annular space. The measurement process starts when the user presses the % H_2O command key on the right-hand side of the panel. First, the instrument's software has a routine to ensure that the cell temperature and the sample temperature are in equilibrium. If they are not, a 'wait' message appears on the meter's LCD until they equilibrate. Then there is a 6 s delay while the microcomputer automatically records the grain's mass, temperature and capacitance, then refers the measured capacitance to the appropriate calibration table in its memory and finally displays the moisture content of the sample, after correcting for its mass and temperature.

The meter has seven commodity keys, accessing calibrations based on ISO and other standards, and it can accumulate up to 254 readings. On command it will display averages in 1 s. Its range is from about 1 to 35% mcwb, with a typical standard deviation of 0.3% mcwb, both dependent on the application and moisture level. It also has a hectolitre weight accessory in the form of a cylinder with an upper slide, shown in Fig. 5.6(b). When the cylinder is filled to the level of the slide and its contents emptied into the meter's cell, hectolitre weight is displayed on command, with a standard deviation of 1 g. A further command key brings up a display of temperature in the range 0–50°C. Temperature is sensed by thermistors. There are other keys which can be used to adjust (*nudge*) calibrations of moisture content and hectolitre weight in 0.1 steps. A security password ensures that only authorised users are able to make these changes.

The meter just described can be calibrated from a master unit or a PC, via an RS232C interface. The master has an identical measuring system but more elaborate software. The moisture content of each sample, determined by a chosen reference method, can be keyed into the unit, or to a PC with installed calibration and communication software (Moisture Net™), which compares the reference method reading with its own corrected reading of the sample's capacitance. From a set of readings at different moisture contents – ideally two to three readings at

0.5% mc steps over the moisture range of interest – it then generates a calibration curve. This can be downloaded to any number of field instruments, to make their readings comparable in subsequent use. The master unit is capable of retaining calibrations of up to 92 different commodities.

Calibrations can also be downloaded over the telephone to field units via a PC with the above software and an intelligent acoustic coupler. At the receiving end the acoustic signals are re-encoded into data (cf the modem in Fig. 2.6) and loaded into the field units. The same software can also interrogate a field unit, update calibration data, perform diagnostic checks and take limited corrective actions.

Summarising, the accuracy attainable with the resistance and capacitance meters depends on the moisture range in which they are being used, the thoroughness with which they have been calibrated against one or other of the oven standards, the accuracy of automatic temperature and weight compensation (where these are incorporated) and the care with which the user performs the measurement according to the manufacturer's instructions. In particular, the user should try to avoid significant differences between the temperature of the meter and that of the samples, since measurement conditions will change until these are in equilibrium, as already indicated. Some manufacturers recommend filling and emptying the cell several times until it has reached equilibrium with the grain temperature. Another temperature-dependent effect is the formation of condensation on the grain or the cell surface, which can occur if a cold tester is brought into a warmer environment. If this happens, unwanted conductivity effects will occur. Measurements on hot grains from a drier are bound to introduce errors, it is better to let the sample cool. This can be easily achieved by spreading the grain on to a small aluminium tray, which will quickly dissipate the heat, or by using a small household fan, thereby speeding up the measurement without incurring too much loss of moisture.

As a general figure, within the normal limits for drying of grain and other seeds and at temperatures from about 0 to 40°C, the best of these meters can achieve accuracies of about $\pm 0.5\%$ mcwb. However, it must not be forgotten that (a) sampling errors are often much higher than temperature errors, and (b) a meter calibrated against one oven standard may appear to be in greater error than this if it is checked against another standard. It is important that the meter should agree closely with the standard that matters when the seed has to be sold.

Grain drying

All forms of farm grain drying employ ventilating air sufficiently dry to reduce the crop's harvest moisture content to its safe storage level (Table 5.2). This must be done as quickly and uniformly as possible, to prevent spoilage, but without adversely affecting the grain in so doing. The two main routes to this end – low-temperature and high-temperature drying – involve different instrumentation and control systems, therefore they will be treated separately.

Low-temperature drying

Here grain is dried in bulk over a period of days rather than hours, using ambient air, which may be assisted by some heating or by dehydration. In order to dry the grain the ventilating air must have a sufficiently low RH, as defined by the crop's moisture content/equilibrium RH relationship. A general form of that relationship for cereals is shown in Fig. 5.7. Since their usual storage moisture content is 14–15% mcwb the RH of the incoming air must fall to about 65% before that level can be achieved. Unfortunately, where weather conditions are like those in the UK there can be long periods when ambient RH is above that level, which can prolong drying and so increase the risk of spoilage. In the UK this condition can occur once in 5 years on average. It can be countered by heating the air sufficiently to reduce its RH to 65% (usually 3 to 5°C) but that brings another risk. If heating is applied too early in the drying process it can cause redeposition of water in the cooler zones of the grain (*capping*) as it passes through them. However, this risk can be avoided by an RH control system developed by the UK Farm Energy Centre. The operator keys in the actual incoming and target grain moisture contents. The controller calculates the required RH to dry the crop, then operates the fan and heaters to achieve optimal drying performance, before initiating a cooling mode to bring back the grain to its storage temperature.

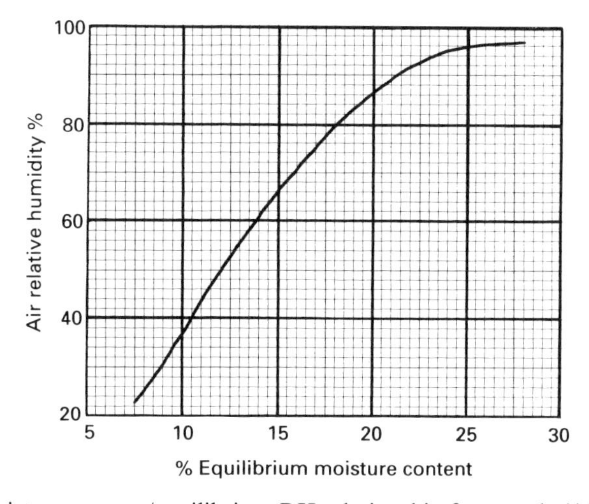

Figure 5.7 Moisture content/equilibrium RH relationship for cereals (ADAS).

The alternative – dehumidification – extracts the moisture from the incoming air by employing the well known principles of the refrigerator. A fan draws air through the cooling coil or *evaporator*, and then through the condenser, where the pumped refrigerant deposits the heat gained from the evaporator. Thus the air temperature is reduced at the evaporator to below its dew point and much of its moisture deposits on the cooling coil, giving up heat to the coil before being collected and drained. The cold, dry air picks up heat again as it passes through the condenser coils and emerges slightly warmer than it was originally, and at much reduced RH. With this method there is a risk of overdrying the lower layers of the grains through which it first passes, therefore it is mixed with some of the untreated air, which bypasses the dehumidifier. The mixing process can be controlled automatically by modulating the bypass flow to achieve the required RH as sensed in the downstream plenum, following mixing by the main ventilation fan. The controller can also switch out the dehumidifier if the ambient RH drops below 65%. This system is most effective when the ambient RH is at its highest, and it reduces both the drying time and the air flow required.

Whether heating or dehumidification are used or not, safe drying at low temperatures requires careful management, based on monitoring of air temperatures and RH, coupled with checks on grain moisture and temperature. Traditional sampling spears are giving way to long instrumented probes, such as the temperature probe already mentioned. Similar probes also measure the RH of the interstitial air, which can be converted to grain moisture content once near-equilibrium conditions are established (and ventilation is off). The older type of RH probe contained a hair hygrometer (section 1.4) but most now operate with an electronic capacitance sensor.

The probe shown in Fig. 5.6(c) combines moisture content and temperature measurement in one unit. Moisture content is measured capacitatively via four conducting fins at the tip of the probe (see fringe-field measurement of soil moisture content, section 3.2). A PRT is embedded in one fin to provide temperature compensation of the capacitance measurement as well as an independent measurement of grain temperature. A microprocessor is housed in the environmentally protected meter, which is shown mounted near the top of the probe. The meter is programmed for six standard crops over the range 0 to 35% mcwb (±0.5% mcwb accuracy). Temperatures can be measured between –20°C and +60°C (±0.1°C accuracy). The probe is supplied in 1 m and 2 m lengths. Both the shaft and the fins can be of stainless steel where hygiene is particularly important. Accessories include an averaging unit and a connector cable which allows the meter to be separated from the probe.

As with the moisture meter shown in Fig. 5.6(b) the meter has a keypad for selecting the crop and command functions; LEDs indicate the active functions and the alphanumeric LCD is back lit for use in poorly lit environments. Calibrations can be *nudged*, wholly or in part, in order to adjust the meter's readings to conformity with a particular reference for an individual crop. The meter's functions can also be changed or extended by changing the EPROM in which the software is resident (see section 2.3, PLCs).

In operation, if the probe is not initially at grain temperature the temperature of the fins will approach equilibrium with that of the grain in the manner shown in Fig. 1.2. Although the fin area helps to reduce this time constant the operator may find it a constraint when the objective is a rapid survey at many points in a bulk of grain. In that event the operator may decide to take readings while the meter's reading is still changing slowly, thereby accepting some loss of accuracy –

directly in the case of temperature measurement and indirectly in the case of moisture content measurement. A difference of 1°C in temperature reading can result in an apparent change in mc of 0.1% mcwb, as stated previously.

The meter's diagnostic features include a visible and audible alarm if the connector cable is in use and a fault develops in it. Cable care is an important responsibility for the user, as stated in Chapter 2. The probe should be kept clean and dry and not exposed to prolonged temperatures above 45°C, while the meter itself should not be subjected to prolonged humid conditions.

High-temperature drying

This category generally refers to driers working at temperatures between 40°C and 100°C and includes batch driers and continuous flow driers. Many of the former and some of the latter control the process by monitoring the temperature of the exhaust air. When the drying air can pick up no further moisture from the grain the exhaust temperature rises to a plateau, determined by the heat losses in the drier. A suitably sited temperature sensor will detect the preset end point on the rising curve and automatically initiate the discharge/reload cycle of a batch drier, or control the throughput of a continuous drier. Other, continuous flow driers employ the more direct method of monitoring the moisture content of the grain via a capacitance measurement.

Figure 5.8(a) shows the form of a capacitance-based controller initially developed in the 1960s for use in intermittent vertical flow driers. The spear-shaped probe is inserted through the wall of the drier near the outlet end of the drying zone, where the bulk density of the grain is fairly uniform. A thermistor is mounted in the probe, to measure grain temperature. The probe head contains a MHz frequency electrical bridge circuit This measures the capacitance of the probe with respect to the surrounding earthed metal, which is dependent on the grain within the enclosure. A control knob on the probe head enables the user to set the target output moisture content of grain from the drier, by reference to calibration tables. Then the moisture signal from the probe represents the error with respect to that set point at any instant. The main control unit applies the requisite corrective action by starting and stopping the drier's unloading mechanism, i.e. the system employs on–off control. In operation the throughput control is inactive until the probe indicates that the moisture content of the grain has fallen below the set point. Then unloading continues until the moisture content rises above the preset differential level. In this way the drying time of the grain is adjusted according to its needs. The system is capable of controlling the output moisture content to within ±0.5% mcwb.

The main limitation of the above measurement is the variability of the bulk density of grain flowing over the probe. One way to overcome this problem is to extract a representative sample of the dried grain from the drier at regular intervals, then to measure its moisture content and temperature automatically in an external cell, where bulk density variations can be brought under control. This still leaves the problem that attends all continuous or semicontinuous flow driers

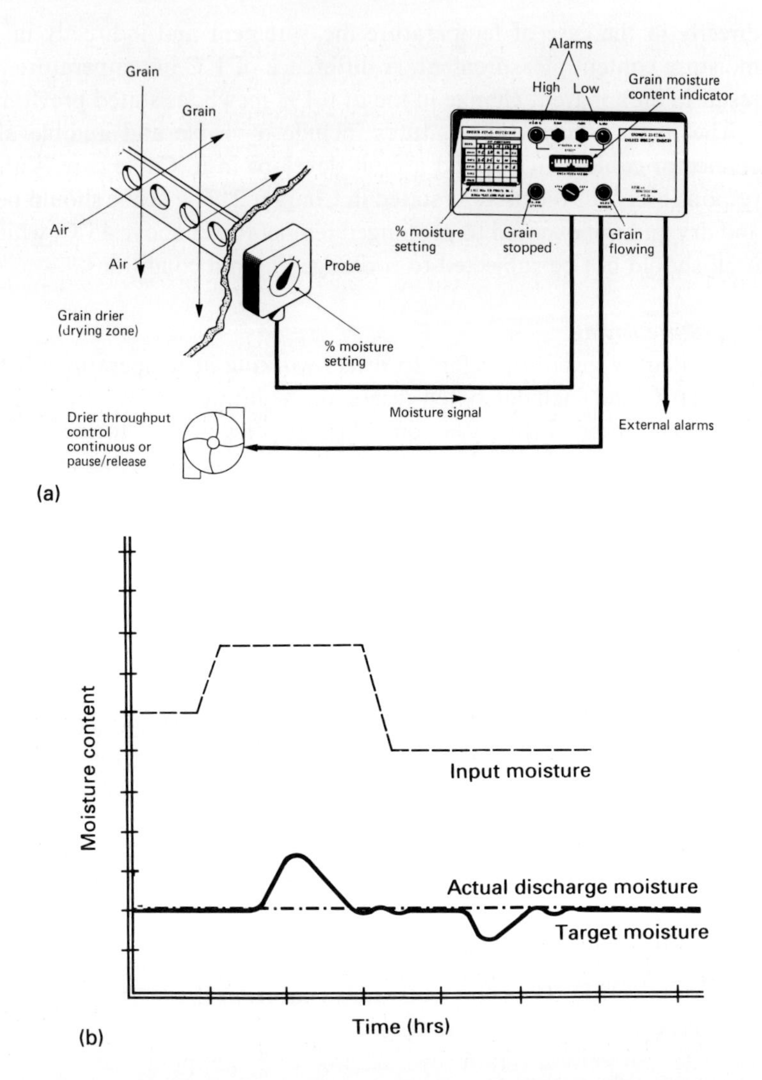

Figure 5.8 Capacitance-based moisture content controllers: (a) schematic (Hunting Electrocontrols); (b) dynamic response (Carrier Bulk Materials Handling).

which rely on a sensor placed at the end of the drying zone. That arrangement cannot anticipate changes in the moisture content of the incoming crop. To overcome this limitation it is necessary to measure the moisture content of the incoming grain automatically and then to apply a control algorithm based on a model of the drying process.

Figure 5.8(b) shows the response of a continuous flow drier with this facility to a sharp increase in the moisture content of the incoming grain, followed by an even sharper and greater decrease. The response at the output is delayed by the drying time of about an hour, and it can be seen that the controller brings the output moisture content back to the target value with a minimum of overshoot and a limited amount of overdrying. Initially, when the drier is filled with a fresh batch of grain, discharge starts slowly, but immediately, and increases until the target output moisture content is attained. That also results in some underdrying at the start but it quickly establishes the steady state. For the most accurate results the control action can be tailored to the crop that is being dried. When the drier is under manual control its grain sampling device continues to function and the discharge moisture content is displayed for the operator's benefit.

Fan control

At this point it should be noted that the soft-start system for induction motors, outlined in section 2.3, is particularly appropriate to fan motors and is another of the energy-saving options in crop drying.

Crop storage

Here it is necessary to distinguish between storage of crops in normal ambient air (99% nitrogen and oxygen, in the ratio 4:1 approximately) and the oxygen-depleted, sealed stores in which crop respiration is greatly inhibited. Both types have the same purpose – to retain the harvest quality of the produce as far as possible until it is marketed – but the cost of the controlled atmosphere (CA) store makes it more appropriate for storage of high value crops such as fruit.

Unsealed stores

Many of these stores are large, with hundreds or thousands of tonnes of crop in bulk or in bins. Instrumentation and control systems are employed for control of temperature, humidity and ventilation, to achieve a uniform environment at temperatures and humidities suitable for long-term storage of a specific crop. Refrigeration and/or humidification may be required to maintain the required temperature and RH. Ventilation is controlled by fans and motorised louvres. Multiple sensor arrays are essential for adequate monitoring of the crop's condition.

The example of potato storage will serve to illustrate the role of instrumentation and control equipment in this sphere. Table 5.4 provides information from which three points can be noted. First, there is only a 6°C gap between the recommended lower storage temperature of ware potatoes and the temperature at which they are liable to frost damage. Second, to avoid moisture loss during storage the air humidity must be high, with the attendant risk that water may be deposited on the crop, thereby creating a disease hazard. Third, from the heat production quoted, the potatoes in a 1000 t store held at 5°C will generate 8 kW of heat. That amount of heat is not readily dissipated, therefore fan ventilation is necessary to prevent a cumulative rise in crop temperature, leading to degradation of its quality.

The simplest form of ventilation is to bring in ambient air sufficiently cool to

Table 5.4 Potato storage data (Farm Electric Centre).

Bulk density	$1.5\,m^3$/tonne	
Loading depth	3.7 to 4.6 m	
Storage temperature	5 to 7°C ware	
	8 to 10°C processing	
Storage humidity (RH)	90 to 95%	
Freezing point (crop)	–2 to 1°C	
Heat emission at	0°C	8 W/t
	10°C	12 W/t
	20°C	18 W/t

restore the crop to its set storage temperature when a temperature rise is detected. To take the example of ware potatoes again, and assuming that 7°C is the chosen storage temperature, effective cooling can be obtained with air at about 5°C and below. The 2°C offset temperature allows for a temperature rise caused by the ventilation fan (about $\frac{1}{2}$°C) and a margin to ensure adequate heat exchange between the air and the crop. However, to avoid frost damage air below 0°C must be excluded. Therefore the periods available for cooling during the storage period (November–March/April in the UK) may not be adequate to keep the crop close to 7°C at all times. Refrigerated air can keep crop temperatures down in milder weather, at additional cost, but air mixing is available as an intermediate measure.

Air mixing can be described by reference to Fig. 5.9(a), which depicts a bulk store with an array of temperature sensors at three levels. Two sets of motorised louvres control the ingress of external air and recirculation of internal air. Internal circulation is often employed to equalise internal conditions, especially when external conditions are not suitable for ventilation. However, for air mixing the two sets of louvres operate together. The initial action is the same as that of the simpler system outlined above. When one or more of the crop sensors registers a temperature above the set storage value, and the ambient temperature is sufficiently low (but above the frost point) the external louvre begins to open, stepwise, to allow in ambient air. However, overriding control is exerted via the duct temperature sensor, downstream of the ventilating fan. This is set to the air mix temperature – again about 2°C below the crop set temperature. The controller then seeks to keep the air duct temperature within a preset differential band, centred on the air mix set point, by step-wise adjustment of the louvre settings. In this way the system makes use of external air that would have become too cold for the simpler system, thereby extending the available periods for natural cooling. The safety thermostat in Fig. 5.9(a) is a further defence against frost damage to the crop.

Additional facilities can be provided by the supervisory control logic. Recirculation routines are initiated when the differential between high-level or low-level sensors exceeds the average sensor temperature by a preset amount. Refrigeration (if installed) can be brought in when external air can no longer be

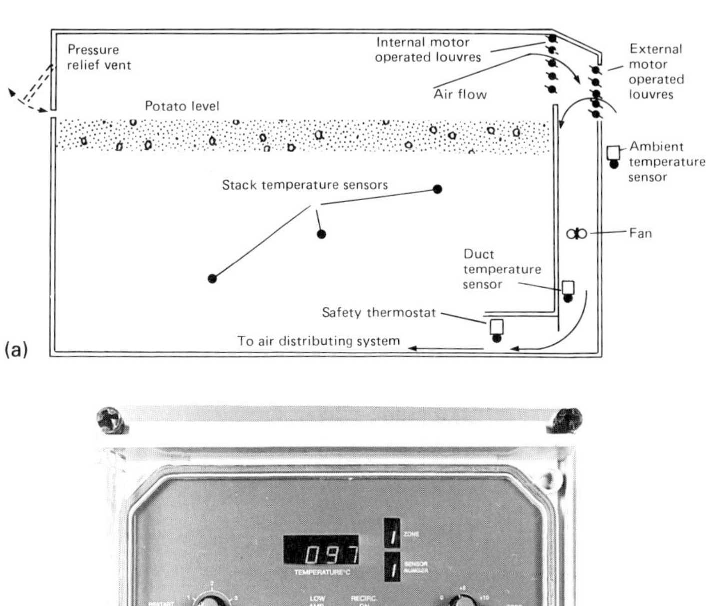

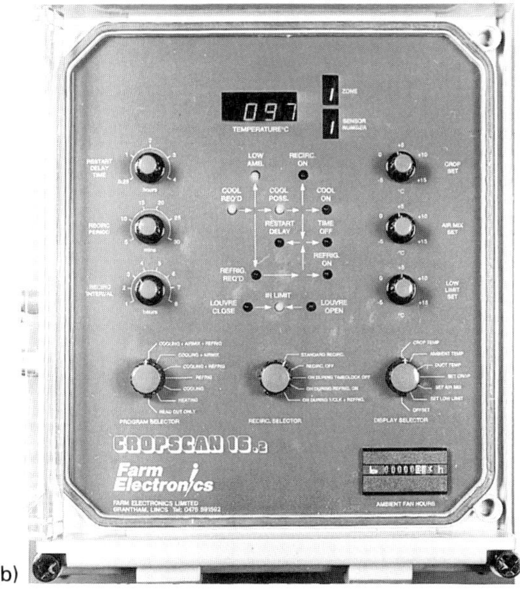

Figure 5.9 Vegetable storage: (a) bulk potato store, schematic (Farm Electric Centre); (b) control unit for a multizone vegetable store, with air-mix facility (Farm Electronics).

used. The monitoring and control panel of such a system is shown in Fig. 5.9(b). This unit can manage several zones sequentially in a large store, each with an array of thermistor temperature sensors – some for crop monitoring and others acting as control sensors. The latter comprise the ambient and air-mix sensors, sensors in the roof space above the crop and low-level sensors in the crop just above floor level. Matched thermistors ensure that a temperature resolution of 0.2°C is attained.

The basic logic is as follows. Ambient cooling takes priority over all other functions and will start automatically whenever a demand is created by any crop sensor, providing:

- The external ambient temperature is at least 2 or 3°C below the highest crop sensor temperature, according to the offset chosen.
- The ambient temperature is not below the preset lower limit.
- The system's overriding time control allows it.

If the ambient temperature is too high refrigeration (if fitted) is initiated. Once initiated, the ventilation or refrigeration will run until all the crop sensors have been brought below the *crop set* temperature or, in the case of ventilation, ambient air temperature no longer satisfies the required conditions. Heating (if fitted) will be initiated if the temperature in store is below a set limit. At the same time the main fan(s) will circulate the warmed air into the roof space and then around the store.

Under air-mix control, if extremely low ambient temperatures inhibit cooling the ambient louvre is closed and the control unit memorises the ambient temperature at that point. When the ambient sensor indicates a rise to 1°C above that point cooling restarts. This is another example of the differential required in on–off control. Recirculation of store air is initiated either on a time control or according to differentials within the crop.

All of these settings feature on the monitoring and control panel in Fig. 5.9(b). Eight of the nine rotary switches are concerned with system settings and the ninth determines which temperature is displayed in the main window. The state of the system at any time can be deduced from the two remaining digital displays and the central array of LEDs. The LED between the *louvre close* and *louvre open* LEDs is lit when the duct temperature is in the differential band (dead band) centred on the air-mix set point.

The electromechanical counter at the bottom RHS of the panel aggregates the running time of the main fan(s), from which important ventilation data can be derived. All data can be transferred via an RS232C link from this store unit to an office PC, which provides graphics and print out of data for management use.

Battery back up is provided for the in-store unit, to maintain programs and control settings during any period of power failure. Diagnostic features include those for sensor integrity. If a sensor or sensor cable fails an out of limits reading is recorded. That reading is ignored unless the sensor is one of the control elements. If the ambient sensor fails the system's fail-safe procedure is to inhibit cooling until the fault has been rectified. In the context of reliability, a particular feature of the system is the design of the ambient louvres. These are made extremely weatherproof and insulated against the large temperature differentials which are typical of this environment.

In systems such as the above, temperature sensors are normally thermistors sealed into the end of long cables, and screened from direct radiation in the case of the external ambient sensor. If humidification is employed water from a set of nozzles can be atomised by the ventilation fan and the resulting RH of the air in the duct sensed by a wet and dry bulb hygrometer. The RH controller can also refer to a similar hygrometer in the incoming air stream, which enables it to calculate the amount of supplementary water required.

Controlled atmosphere stores

Long-term storage of apples (specifically Cox's Orange Pippins) at constant low temperature in a sealed, low-oxygen/high-carbon-dioxide environment has a 60-year history. Oxygen is naturally depleted and carbon dioxide increased by respiration of the fruit. When carbon dioxide is *scrubbed* from the atmosphere continuously, by circulating it through activated carbon, the concentration of this gas can be maintained at under 1% (10 000 vpm) – i.e. below about 30 times its normal atmospheric levels. Oxygen can be reduced to below 2% (less than a tenth of its normal level). The remainder of the air is mainly nitrogen. However, below 2% oxygen there is an increasing risk that alcohol will be generated by the fruit, with resulting loss of quality, and therefore control systems for ultra-low oxygen have been developed. These are based on the determination of oxygen concentration by paramagnetic analysers (section 1.16). Carbon dioxide is measured by the infra-red gas analyser which has been introduced in section 1.16 and elsewhere.

Therefore management of the environment in ultra-low oxygen (ULO) stores requires precise control of temperature (at just over 3°C), oxygen and carbon dioxide, using refrigeration and scrubbing equipment for regulation of the first and last quantities respectively. Oxygen levels are adjusted, as necessary, by controlled admission of external air. The adjustments can be made manually but the requirements are stringent, the cost of error high and the task very time-consuming, especially when there are several stores in an installation. Here, automatic control is not only time-saving but also more likely to achieve the environmental regime that is required.

A modular system for CA monitoring and control in up to 62 stores is shown in Fig. 5.10. The computer employs a numeric membrane keyboard for simple, step-by-step entry of data and control instructions. Data and operational information are displayed by an LCD panel. All computer interfaces employ optoelectronic isolators which convert signals from electrical to equivalent optical signals and back again as in the PLC (section 2.3). These protect the computer from hazards due to noise on the signal lines. Internal battery back up maintains the computer memory and clock for at least 4 hours in the event of a power failure. The measurement cycle restarts automatically if the power is restored within that time. Provision is made for connection of an external battery and charging unit if a longer period of protection is required. Operating programs can also be transferred to a removable EPROM, for long-term storage until they are needed again. Serial ports are available for a printer and remote terminals.

The switching module immediately below the computer in Fig. 5.10 allows the user to select automatic operation or to operate the system manually. Next, in descending order, are a CO_2 meter (0 to 15% linearised output) and its associated

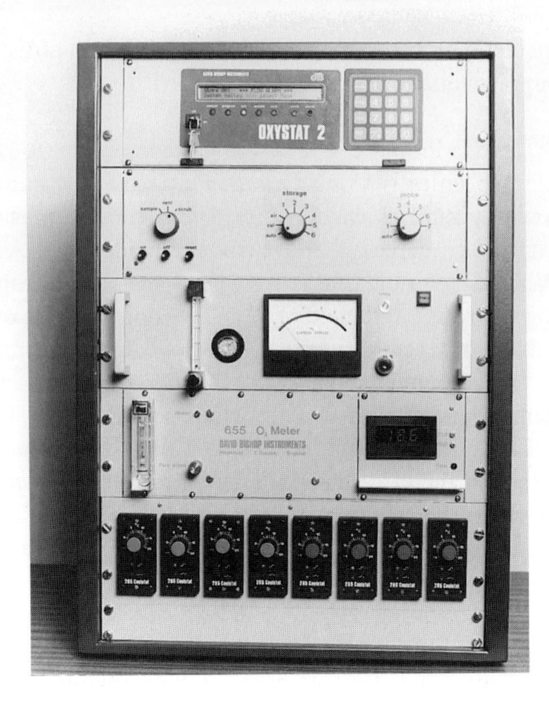

Figure 5.10 Controlled atmosphere storage: monitoring and control unit (David Bishop Instruments).

flow meter; temperature-controlled dumb-bell type O_2 meter (0 to 21%) with its flow meter, and a row of temperature control modules (–5 to + 20°C setting).

In operation, the store atmospheres are sampled at preset intervals between 1 and 999 min and for preset durations between 1 and 9 min. The real time of all measurements is displayed as they occur and is recorded with them. Recording intervals are preset in the range 1 to 999 measuring cycles. The same range is available for the introduction of calibrating gases, which are employed to correct subsequent readings of the oxygen and CO_2 analysers. For each store the user can program in oxygen, CO_2 and temperature set points, high and low alarm levels for each of these measured quantities, ventilation time when there is a need to increase oxygen levels and scrub time for the depletion of CO_2. The actual ventilation and scrub times can be multiples of the programmed times, depending on the size of the error.

The peripheral units include the sampling equipment, which is based on a high-quality pump and solenoid valves, operating in sequence to extract gas from each store in turn. This equipment also contains a vacuum gauge for leak testing. The calibrating gases are introduced through the sampling system at the preset intervals. A 95% nitrogen: 5% CO_2 mixture is employed to check the working range of the CO_2 meter and the zero of the oxygen meter. Fresh air checks the 21% reading of the latter and the near-zero response of the former. All samples are exhausted to the outside air after passing through the analysers.

The analysers are part of the central installation, of course. Store temperatures are measured by thermistors (up to seven per store: maximum error ±0.2°C over the range –25 to +25°C). Measurements of water loss in stores can be added via transducers in their drainage system, to enable calculation of the weight lost by the stored crop.

The hard-to-store Cox's Orange Pippin can retain its eating qualities for up to 6 months by this means of environmental control. Other high value fruit and vegetables are being stored in the same way – each requiring specific and closely defined control of the environment. Controlled atmosphere regimes of 3% O_2 and 6% CO_2, at 5°C storage temperatures, have also been used to regulate the sprouting of potatoes.

5.4 Crop quality determination

Instrumentation and control systems for crop quality determination mostly follow harvesting, taking their place in cleaning and grading installations. Therefore, automatic methods must be demonstrably equal to – if not better than – human graders in making consistent quality decisions. They must also provide the required throughput without damaging the produce. These considerations appear to rule out many of the experimental techniques that have been devised. Here some well established methods are reviewed, together with newer techniques which appear to show promise.

Optical methods

The most successful methods by far have been based on the body reflectance of the produce, mentioned in section 4.3. Figure 5.11 illustrates a method for high-speed sorting of seeds of many kinds, first developed in the late 1940s. Its purpose is to remove discoloured seeds from a stream of the product, fed from a V-shaped chute. These machines now employ bichromatic colour sorting which can distinguish variations in hue and saturation (see Fig. 4.8), thereby making it possible to separate seeds with quite fine colour distinctions. The system is also used with processed products, such as dried, hydrated vegetables, under computer control, which compensates for slow changes in the illumination. Up to nine images of each item can be processed by reference to a stored calibration for the specific product, at a throughput of the order 100 kg/h/channel. Conveyor belt sorting of harvested fruit and vegetables has been developed to employ the same bichromatic system. Rejection of unripe produce, stones, soil and other unwanted material is effected by fingers similar to those employed by the X-ray sorter shown in Fig. 4.4.

Colour sorting of citrus fruits and tomatoes is also well established. Bichromatic measurements on the latter crop make it possible to divide them into two or three colour grades, each covering a colour range that can be modified by the user, to suit changing market requirements. Typically, the tomatoes are fed into individual cups in a multichannel conveyor, which carries them first into the

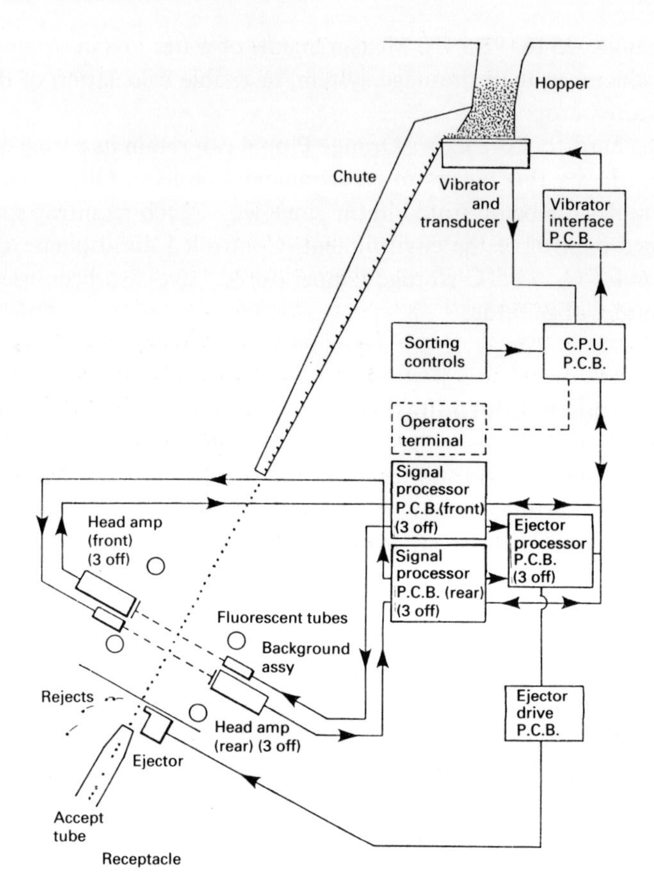

Figure 5.11 High-speed sorting of seeds, schematic (Sortex).

viewing enclosure, illuminated by halogen lamps. Two colour cameras or photodetectors colour grade the visible contents of each cup and pass the information to a central computer. Then the hinged cups pass a weighing section, where the free end of each one is briefly lifted from its conveyor support and its weight supported by a load cell. Since the tare weight of the cups is known the load cell's output can be offset, to register the weight of the cell's contents. The computer combines the colour and weight information received from each cup and then allocates the tomato in it into one of the preselected colour/weight grades. The computer also calculates how far each cup has to travel to the appropriate delivery point and employs a conveyor motion sensor to determine when the cup has reached its destination. At that point it actuates a mechanism which releases the cup's conveyor support and the cup drops to release its contents into the selected colour and weight channel. The support is restored and the stored information on the discharged tomato's weight and colour grades is transferred to the computer's data store on disk or other bulk memory device. The cup can then be reloaded at the filling point and the cycle repeated. The data store is used to provide summarised information on tomato numbers/grade and so on. A ten-lane system of this type can colour/weigh grade over 100 000 tomatoes/h.

Machine vision

Taken broadly, this term (or the alternative, computer vision) can apply to any system, optical or otherwise, which provides a computer with an image of a scene from which it can extract significant features of one kind or another. In practice, most applications up to now have been based on images taken at optical wavelengths or the near infra-red. Examples have already been given in Chapter 4, in the context of the autonomous vehicle and the robotic harvesting of citrus fruit (sections 4.2 and 4.3). The information is gathered by the CCD camera, introduced in section 1.17, coupled with the frame grabber which provides a succession of frozen images. From that point development is mainly concerned with the application of suitable algorithms to locate features in the scene, coupled with artificial intelligence for decision making.

Onc of the first advantages conferred by image analysis is that it can free a grading line from the necessity of separating objects before classifying them by shape and size. This capability has been deployed on a potato grading line which uses an overhead camera stationed above a roller conveyor, as in Fig. 5.12(a). Figure 5.12(b) shows an image generated by the computer, with tubers of different shapes and sizes, touching in places and in one case overlapping slightly. The stationary bed beneath the rollers causes them – and the potatoes – to rotate as they pass through the *firing lines*. This gives the computer the opportunity to collect all-round images of the tubers, from which it can deduce sizes and shapes, and estimate weights from calculated volumes.

However, before it can complete that task it must separate apparently connected tubers into individuals. First, the outlines shown in Fig. 12(b) are obtained

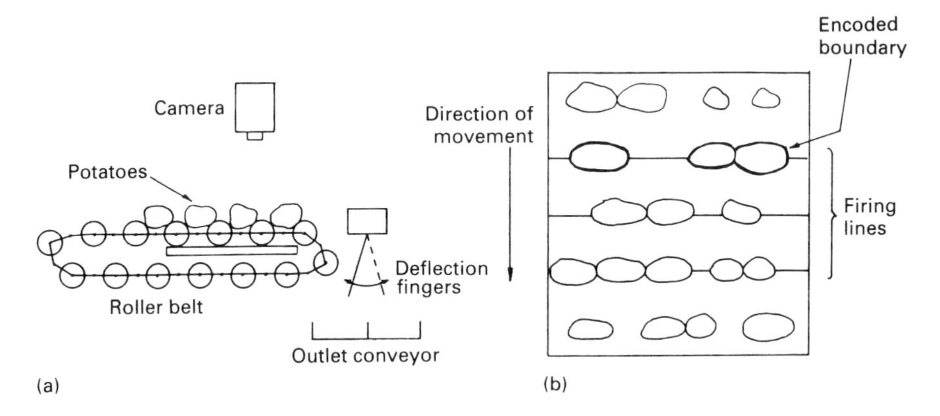

Figure 5.12 Potato grading by machine vision: (a) system, schematic; (b) computer imaging: potato shapes and positions (Silsoe Research Institute).

by a process known as chain coding of the image pixels, which identifies the tubers' boundaries. That is done in the firing lines, and thereafter only the boundary data are required for further processing. The computer's most difficult task is to decide whether there is one tuber or two where boundaries overlap, as they do in the case of the tubers at the right of the topmost firing line. It resolves that issue through its search for cusps in outlines of continuous boundaries. Here decision making calls for rules governing the identification of a genuine division – e.g. that there must be a pair of cusps, with one on the upper boundary and the other on the lower boundary and no farther apart than a prescribed vertical or horizontal distance. Multiple images of the same pair of tubers as they roll through the camera's field of vision give the computer the opportunity to make several judgements of the tubers' unity or separateness. After this and other processing the computer makes a further decision on each tuber's grade, based on the accumulated data, then ensures that it is deposited in the correct collector, downstream, by reference to the conveyor's speed.

Figure 5.13 shows a related activity – determining the outline of growing mushrooms, in order to measure their size and position. This operation is the first stage for selective harvesting of mushrooms robotically, now in an advanced stage of development. The outlines shown were produced by an edge-tracking algorithm which progressively explores the grey-scale image pixels produced by a black and white camera. This is not identical to the chain-coding algorithm, although it produces a similar result. Individual mushrooms can be picked out because they are diffuse (matt or Lambertian) reflectors, which make their top-lit centres their brightest parts. Therefore, high threshold segmenting finds their

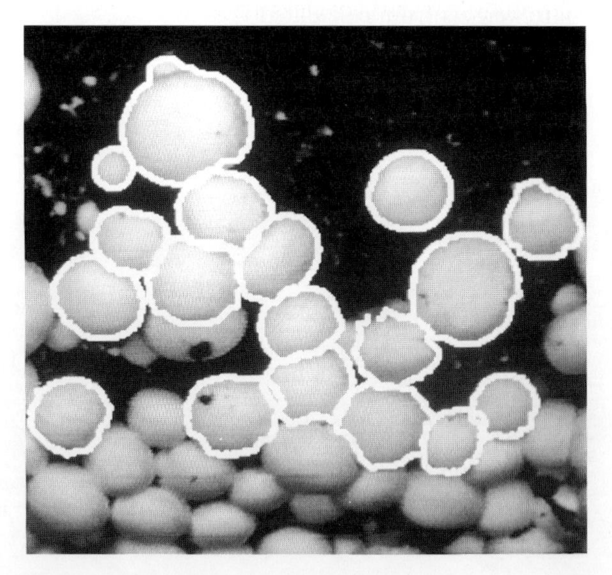

Figure 5.13 Machine vision: location of mushrooms for harvesting (Silsoe Research Institute/University of Wales, Cardiff).

approximate centres, from which points a search can be made for their outlines using a low threshold to provide good segmentation between mushrooms and background. Faced with an image such as this, however, the harvester needs another algorithm not only to decide which mushrooms to pick, but in which order, since there is both close contact and considerable overlapping of their images. These algorithms have been developed, together with a harvesting arm with the necessary gentle action.

Automatic detection and sizing of visible blemishes on fruit is another area of development, since the size of a blemish is a factor affecting the quality grade of a fruit. This also requires algorithms which can locate the boundaries of features in an image of the fruit. Several methods of doing this have been developed.

Also in the horticultural sector automatic grading of pot plants by size, shape and colour can be achieved with an adaptable commercial system developed at the Netherlands Agrotechnological Research Institute. This is shown diagrammatically in Fig. 5.14(a). Feature extraction is followed by classification, in the usual way. The first stage is to select colour groups representing features of the plants to be sorted. In the case of flowering plants, there can be seven groups, representing background, leaves and five colours. The colour recognition module can be trained to recognise these groups by presenting it with representative images. The neural network employed for colour classification of Saintpaulia plants is shown in Fig. 5.14(b). After training, the system has been shown to have a false colour recognition rate of less than 0.1% in operation. This represents fewer than three misclassified plants per hour at normal operational speeds of 3000 plants per hour. The measurement modules of Fig. 5.14(a) are additional geometric features such as size, width and diameter, including shape and symmetry descriptors. The ensuing class assignment, based on the extracted features, is learned either by the neural network route, as with colour recognition, or by

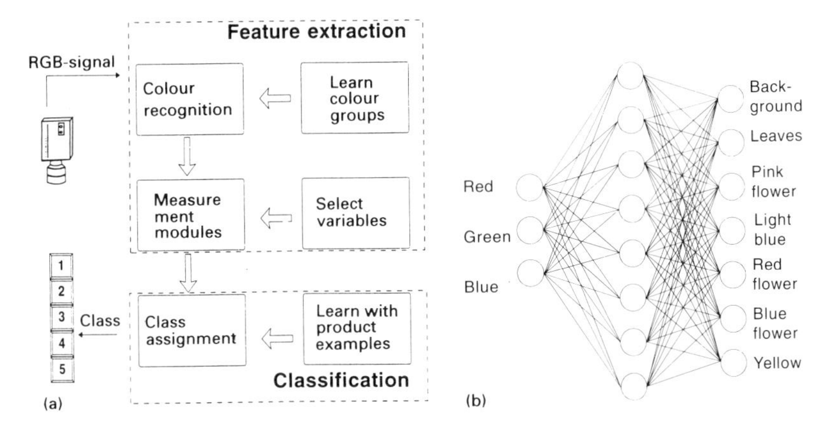

Figure 5.14 On-line sorting of plants: (a) classification method; (b) neural network structure (ATO–DLO, The Netherlands).

statistical techniques. Training sets are selected by experts in the individual product area (e.g. cactus plants).

Internal imaging

Many techniques have been employed to detect internal defects in agricultural produce, with particular interest in fruit. The general problem with most of them is that they are not inherently suitable for real-time, on-line assessment at speeds consistent with grading line operations. Many, too, are only capable of detecting gross defects in the product. A possible exception is the application of nuclear magnetic resonance (NMR), which is widely used in the food industry for rapid analysis of moisture, oils and fats. Its use with agricultural produce dates back to the early 1950s. An outline of the NMR method was given in section 1.15 and its use for moisture determination in soils is included in section 3.2.

Rapid analysis requires the pulsed RF method rather than the use of continuous wave (CW) energisation of the samples. A sharp 90° RF pulse is injected into the sample by the external coils, causing responsive nuclei to precess round the direction of the magnetic field at an angle approaching 90°. The pulse

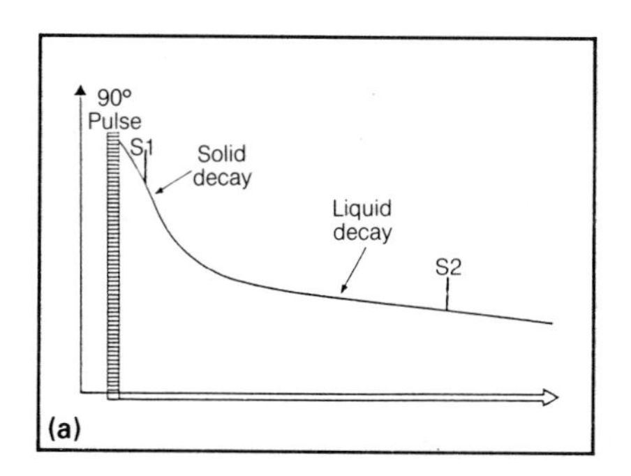

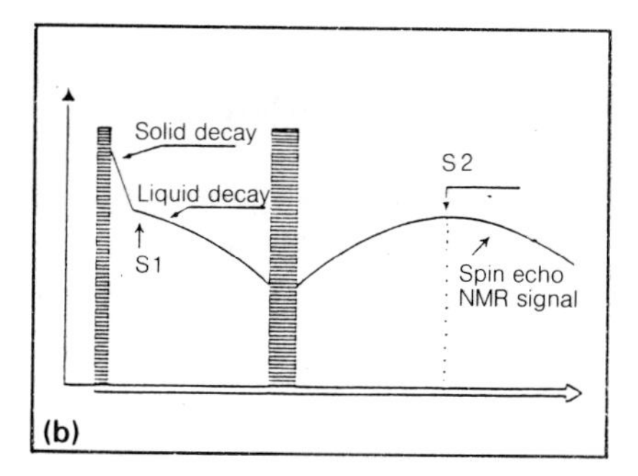

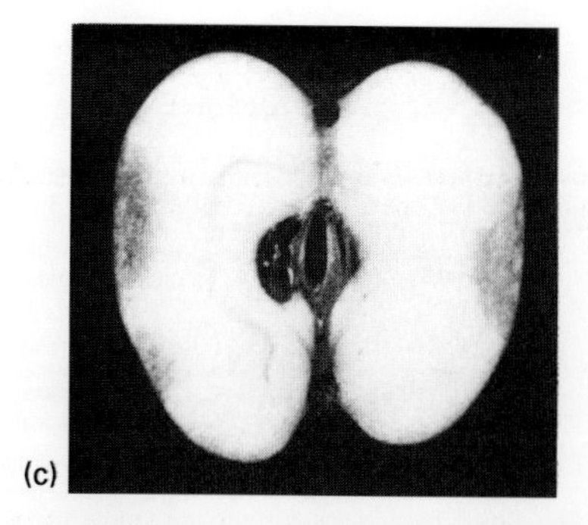

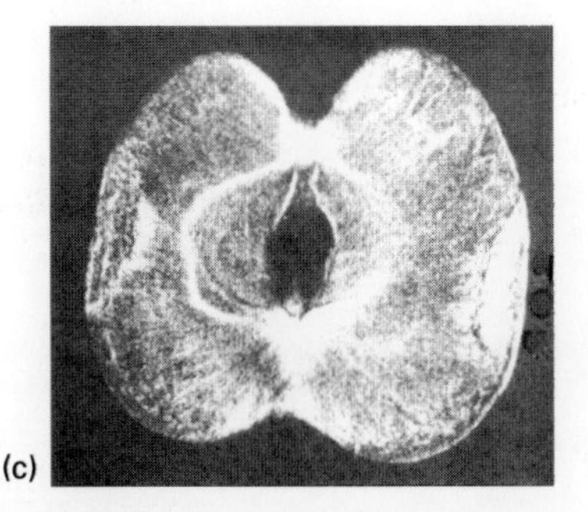

Figure 5.15 Internal imaging by pulsed NMR: (a) single pulse, and (b) spin-echo technique (Bruker); (c) image of bruised apple tissue (right), compared with the actual bruises (University of California, Davis).

duration is only a few microseconds. Subsequently the nuclei return to their original state in relaxation times dependent on their physical and chemical status. The profile of the return RF signal is then in the form shown in Fig. 5.15(a), where the peak response is determined by the abundance of all the responsive nuclei and the decay is a composite of all their relaxation times. It can be seen that the precise abundance of component S2 (which could be an oil) is difficult to determine from the curve. Therefore, the spin-echo method was developed. That follows the RF pulse by another, broad-band pulse, termed the 180° pulse, at a time T (usually a few ms later). At time $2T$ that produces a spin echo, as shown in Fig. 5.15(b). By that means the abundance of S2 can be more easily calculated.

Magnetic resonance imaging (MRI) adds another feature. As stated in section 3.2, additional gradient coils are employed to impart gradients to the magnetic field. In consequence, the RF frequency at which a given nucleus resonates will depend on its position in the magnetic field. That makes it possible to build up a two-dimensional image of the abundance of a nucleus in a slice of the sample – and ultimately to build up three-dimensional images. Magnetic resonance imaging instruments can now generate a two-dimensional image in less than 100 ms.

Figure 5.15(c) shows an MRI image of the centre section of an apple alongside a subsequent photograph of the same apple cut through the scanning plane.

5.5 Further information

Published papers which have provided examples of techniques described in this chapter are listed below, with others of a more general nature.

Section 5.2

Herold, B., Truppel, I., Siering, G. & Geyer, M. (1996) A pressure measuring sphere for monitoring handling of fruit and vegetables. *Computers and Electronics in Agriculture*, **15**, 73–88.

Section 5.3

Matthews, J. (1963) An automatic moisture content control for grain driers. *Journal of Agricultural Engineering Research*, **8**, 207–20.

Section 5.4

Machine vision

Marchant, J.A. & Sistler, F.E. (eds) (1993) Special issue: computer vision. *Computers and Electronics in Agriculture*, **9**, 1–120.

Marchant, J.A., Onyango, C.M. & Street, M.J. (1990) Computer vision for potato inspection without singulation. *Computers and Electronics in Agriculture*, **4**, 235–44.

Tillett, R.D. & Batchelor, B.G. (1991) An algorithm for locating mushrooms in a growing bed. *Computers and Electronics in Agriculture*, **6**, 191–200.
Timmermans, A.J.M. & Hulzebosch, A.A. (1996) Computer vision system for on-line sorting of pot plants using an artificial neural network classifier. *Computers and Electronics in Agriculture*, **15**, 41–55.
Yang, Q. & Marchant, J.A. (1996) Accurate blemish detection with active contour models. *Computers and Electronics in Agriculture*, **14**, 77–89.

Internal imaging

Callaghan, P.T. (1991) *Principles of Nuclear Magnetic Resonance Microscopy*. Oxford University Press, Oxford.
Chen, P., McCarthy, M.J. & Kanten, R. (1989) NMR for quality evaluation of fruits and vegetables. *Transactions of the American Society of Agricultural Engineers*, **32**, 1749–53.
Zion, B., Chen, P. & McCarthy, M.J. (1995) Detection of bruises in magnetic resonance images of apples. *Computers and Electronics in Agriculture*, **13**, 289–99.

General

Bull, C.R. (1993) A review of sensing techniques which could be used to generate images of agricultural and food materials. *Computers and Electronics in Agriculture*, **8**, 1–29.
Chen, P. & Sun, Z. (1991) A review of non-destructive methods for quality evaluation and sorting of agricultural products. *Journal of Agricultural Engineering Research*, **49**, 85–98.
Tillett, R.D. (1991) Image analysis for agricultural processes: a review of potential opportunities. *Journal of Agricultural Engineering Research*, **50**, 247–58.

Chapter 6
Livestock Production

6.1 Introduction

This chapter covers the requirements and performance of livestock of different species, reared for different purposes and providing a variety of end products. As with crop production, instrumentation and control systems can provide data needed for better informed decision making, labour saving and improved process control in all of these spheres. In addition they can contribute to the welfare of farm animals in important ways.

Inevitably, the chapter is dominated by developments for the dairy herd, since milk producers adopted automation as herd sizes increased, starting with large Californian units in the late 1960s. Nevertheless, recently developed technologies such as machine vision and robotics are beginning to make contributions in other parts of the livestock sector.

6.2 Livestock environment

The most common measurements in this context are temperature, RH and air movement (in that order), although there is a requirement for detection of pollutants such as ammonia and CO_2. For naturally ventilated buildings, hand-held instruments are available from many suppliers for checks and surveys of interior conditions. Automatic control of natural ventilation (ACNV) is also available. This employs motor-driven side ventilators, coupled with motorised roof baffles in some cases, which open and close, stepwise in response to changes from the desired temperature set point. The sensors are normally one or more (preferably more) thermistors. Temperature rises are due to solar heat gain and livestock heat emission, of course: supplementary heating will only be brought in at minimum ventilation settings.

Forced ventilation by fans provides a more positive means to distribute air within the building, in ways that increase control of the animals' environment. Figure 6.1(a) shows a system developed for that purpose at the NIAE (now SRI) in the 1970s. Research had shown that very stable air circulation patterns could be obtained in a pig or poultry house, over a wide range of external conditions day and night, as well as winter and summer. The means of achieving

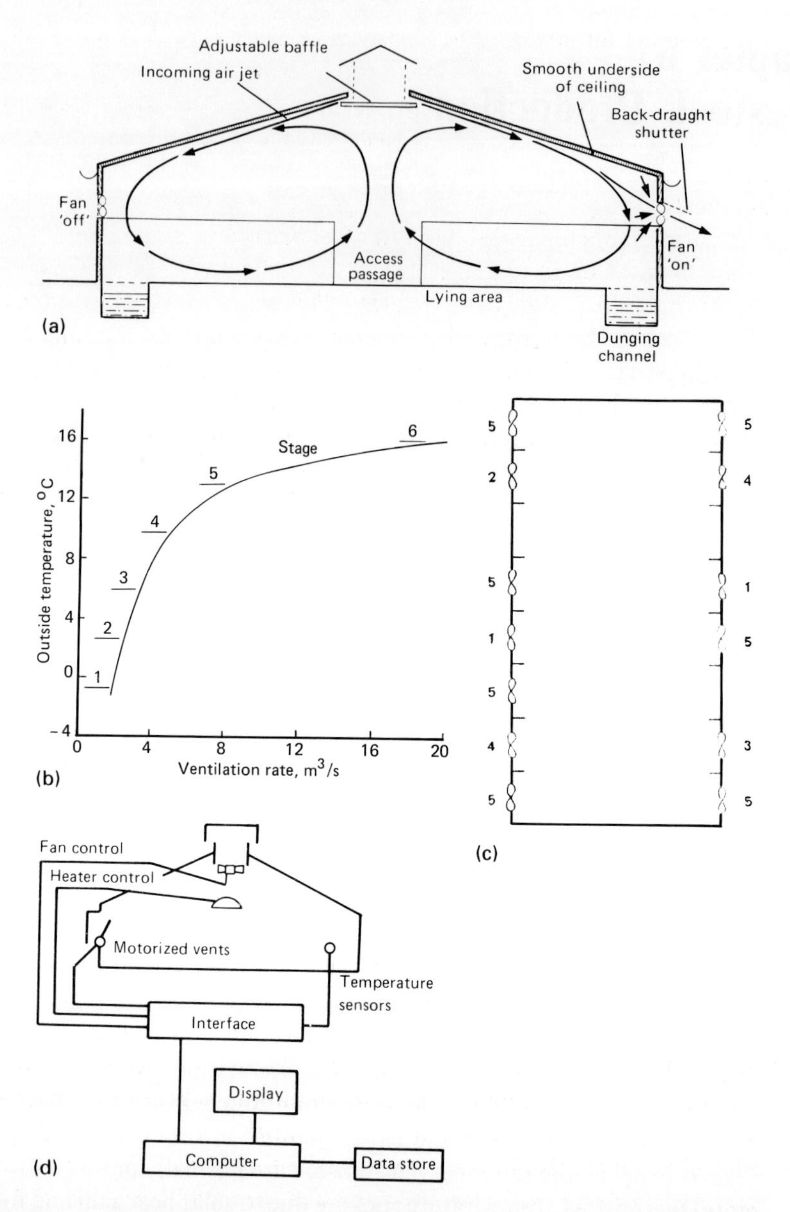

Figure 6.1 Automatic control of air temperature, using step control of fans and automatic vents: (a) section of a piggery with a central feeding passage and side dunging channels; (b) six-stage fan settings, covering a 10:1 range of ventilation rates; (c) fan switching sequence; (d) control system for a poultry house (Silsoe Research Institute).

this were very simple, in principle. Referring to Fig. 6.1(a), the system required an adjustable inlet baffle, running along the ridge of the roof, together with a set of extractor fans, mounted in each side wall. These two elements of the control system worked together to produce the stable circulatory air currents shown in the diagram, given an undersurface to the roof which did not impede the incoming air stream. The design objective was to obtain an inlet air speed of about 5 m/s at all times, regardless of the total volume of air passing through the house in m^3/s. This was achieved by a control algorithm which switched the fans on or off in sequence as the internal temperature rose or fell, respectively, and simultaneously opened or closed the roof baffle by preset amounts.

Figure 6.1(b) shows the calculated ventilation rate for a particular building, over a range of ambient temperatures. The curve was obtained by first calculating the maximum required ventilation rate. This was done by multiplying the number of livestock to be housed, their average weight and the recommended rate/kg liveweight. The minimum ventilation rate was then set at one-tenth of the maximum. The thermal characteristics of the building were introduced, to take account of heat transfer, and the relationship between ambient and internal temperatures and ventilation rate calculated.

The calculated step-wise control of the fans is also shown in Fig. 6.1(b). The on–off form of control was chosen because maximal fan efficiency is achieved in this way and because the performance of variable-speed fans can be affected by wind when they are running at low speeds. Time delays ensured that stepping from one stage to another could not occur at less than 3-minute intervals, thereby allowing time for conditions to stabilise at each stage. The number of fan stages employed in this system is not crucial but five or six stages are preferred. These should be staged at equal ambient intervals, to produce a similar cooling capacity per stage, at 3 to 5°C steps. In order to avoid uneven air distribution along the house, the fans should also be switched in groups, as shown by the example in Fig. 6.1(c). The curve in Fig. 6.1(b) made it possible to calculate the roof aperture required at each stage in order to maintain the 5 m/s air flow already referred to, while allowing in enough air to maintain the set internal temperature. This temperature was monitored by connected groups of matched thermistors, to provide an average reading, since the output of a single sensor can be unrepresentative, particularly if it is sited in a building with variations in stocking density. Light back-draught shutters, mounted externally and hinged at the top, were employed to seal off each fan aperture when the fan was not running. In its closed position each shutter was held in place by a small permanent magnet, mounted in the wall aperture, which attracted an equally small steel block, fixed to the lower end of the shutter. When the fan was switched on, the force of the air overcame the magnetic attraction and the shutter swung out horizontally, thereby offering very little resistance to the flow. The fail-safe facility needed to deal with a power failure or uncontrollably high temperatures was provided by a simple electromagnet system. Either of the two preceding conditions de-energised the electromagnets, which released (a) a spare loop of cable in the mechanical linkage to the adjustable baffle, and (b) a weight attached by another cable to the shutters. This dropped the baffle and pulled the shutters to their fully open positions. Systems of this type have been installed in finishing piggeries, broiler and turkey houses, where they have achieved close control of the environment.

Poultry environment

The temperatures required for maximum production from adult birds lie in the range 13 to 16°C for broilers and 21 to 24°C for layers, according to authorities in this field. Air flow should be low to avoid chilling the birds and RH should lie in the 60 to 75% range for health reasons. Hazardous levels of gases such as carbon dioxide and ammonia (both generated by the animals) must be avoided, too. Day-old chicks require local temperatures of about 30°C, reducing daily by 0.5 to 1°C over 3 weeks to the background house temperature of about 20°C.

For broilers a variant of the NIAE system, shown diagrammatically in Fig. 6.1(d), has been developed. This reverses the original procedure by installing the fans in the roof and employing motorised vents in the side walls. Heating is also incorporated, with stirring fans, to maintain the required internal temperature in cold weather, or at night, while allowing the system to maintain the preset minimum ventilation rate. The system also incorporates a computer which automatically adjusts temperature and ventilation rates on a daily basis, as the birds grow. The required temperatures are based on recommended levels, and the minimum air flows are calculated in the manner already outlined. The basic information is stored in look-up tables in the computer's memory. A real-time clock enables the computer to make the above adjustments at the required times and battery back up makes it independent of mains power. Alarm indicators to warn of breakdown of the system are also included. Once the user has switched the system on, entered the day number and the number of birds, operation of the *start* button initiates the program of automatic control.

Not every farm building lends itself to the step-control, automatic vent system, of course, and many other fan-ventilation controllers are available to suit different requirements. Some employ variable-speed fans, combined with proportional control action (i.e. the fan speed is proportional to the difference between the set point and the actual signal derived from a temperature sensor or a group of sensors). The sensor is usually a thermistor. Since the ventilation must never be zero, it is usual to set the minimum fan speed at about one-tenth of its full speed. The controller is adjusted to provide proportional action between these two speed limits. The resulting temperature control band (the *modulating* band) is illustrated in Fig. 6.2(a). The diagram only depicts the controller's action, of course. Proportional control has the limitations outlined in section 2.3 and the actual temperatures achieved will depend on the design of the complete system. Design factors such as building shape and layout, number and location of fans, efficiency of fans at reduced speed, design of air inlets and outlets, and the location of the sensors have an important bearing on its overall performance. However, as in all control systems, the performance will deteriorate if the sensors are not properly maintained.

Where both heating and ventilation system are installed in the house the controller can modulate the fans above the temperature set point and modulate the heater below it, as shown in Fig. 6.2(b). This avoids wasteful competition

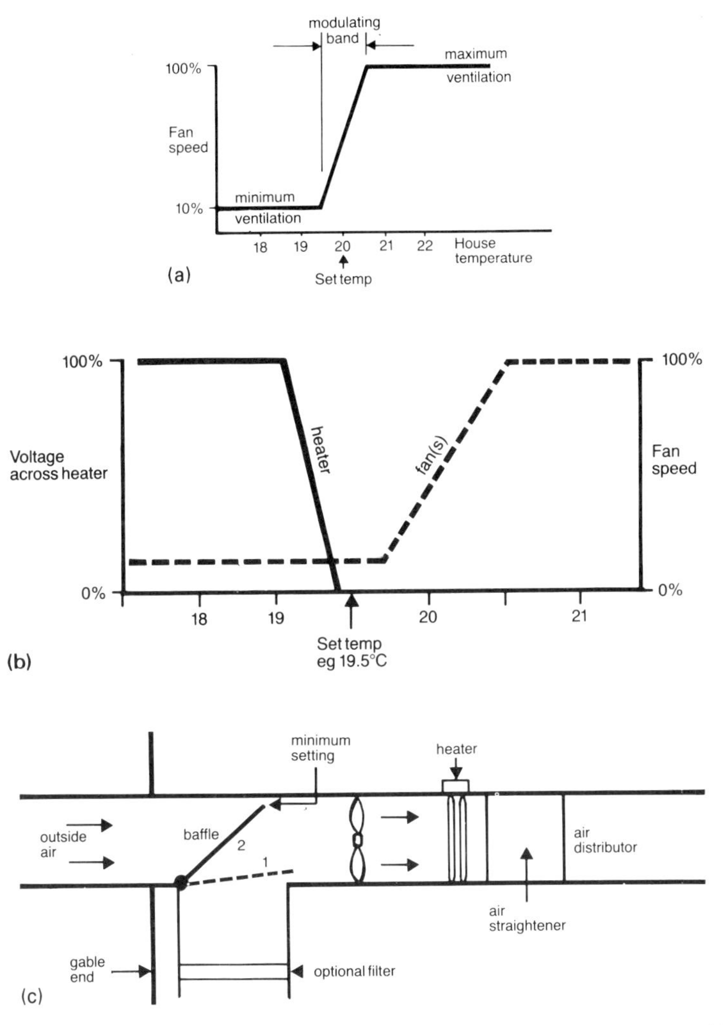

Figure 6.2 Automatic control of air temperature, using proportional control of fans: (a) modulating band; (b) combined heating and ventilation; (c) air recirculation (Farm Electric Centre).

between the two systems near the set point. Time-proportional control has been employed for this purpose in poultry houses, since it is compatible with digital equipment. In this modified form of on–off control the percentage *on* time in a fixed cycle increases as the temperature error increases.

One problem that can be experienced with proportional control in this context occurs at start up, when the fans can be running in reverse, due to the effect of strong wind. If the house temperature is near the set point at that time the controller may not provide enough power to drive the fans in the required direction and this creates a risk of damage. Therefore, a short initial *boost start* (full-power) is sometimes applied by the controller to ensure that any reverse rotation is quickly overcome.

Some controllers operate with recirculation systems, such as that shown in Fig. 6.2(c). In this case they have to control the angle of the baffle which regulates the amount of outside air brought into the house.

The special requirements of chicks are met by brooder controllers which operate with infra-red heaters (electricity or gas-powered) to achieve the gradually changing temperature regime specified for these birds. The sensing element in the gas-fired heater shown in Fig. 6.3 is a *black body* temperature sensor, which absorbs a wide band of infra-red wavelengths. This is intended to simulate the absorption characteristics of the chicks and so to provide a measure of their skin temperature. In use it is suspended at about chick height, where it can provide a representative reading of the radiation received. Its internal temperature sensor provides the necessary input to the energy regulator. A single sensor and controller can manage an array of brooders, thereby providing energy economy at low capital cost.

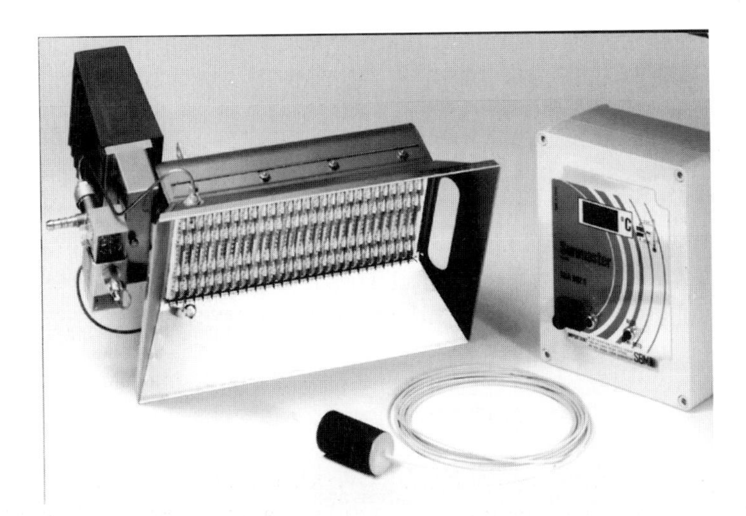

Figure 6.3 Gas-fired radiant heater for a poultry brooder, with electric ignition, flame-out safety switch and black-body temperature sensor (SBM International).

Hatcheries

Chick hatching in incubators as a large-scale operation is closer to factory production of farm supplies than to farm production itself. Nevertheless, it is a process requiring precise control of the environment in which the egg's embryo develops between on-farm laying and the return of the newly hatched chicks to the farm environment. It thereby closes a large loop, involving environmental control at all stages. Since hatchability can change by a much as 10%/°C, control of temperature at all stages of the process must be as accurate as possible. If eggs have to wait in storage, temperatures are not so critical (13 to 16°C is common) but once in the incubator their temperature must approximate to 38°C. Relative humidity levels are also important, in relation to water transfer between the egg and its environment, although they are not so precisely defined. During storage RH is maintained at around 75%, but this reduces to between 60 and 50% during incubation, until the final stages of the 3-week process. At that point the RH is first reduced, then increased at hatching and finally reduced to allow the newly hatched chicks to dry off. Oxygen and carbon dioxide levels are equally important to the embryo's growth. Oxygen concentration must not drop below 15% and carbon dioxide levels should lie between 0.1 and 0.4%. Carbon dioxide affects the alkalinity of the albumen, which should lie in the range pH 7.2 to pH 7.4.

Pig environment

The air temperature requirement for the most efficient production of pigs for meat is at least as stringent as that for poultry. The pig has a well attested thermoneutral zone, which can be expressed mathematically in terms of air temperature, air flow over the animal, liveweight, number of pigs in a pen, metabolisable energy (ME) value of the animal's food and the thermal properties of the pen's flooring. The last-mentioned factor varies significantly between concrete, wood and straw bedding. From the relationship of these quantities a lower critical temperature (LCT) can be defined, below which the pig increasingly utilises feed energy to maintain its body temperature, thereby wasting the costliest input in pig production. The relationship for fattening pigs in the 40 to 70 kg range can be seen graphically in Fig. 6.4, which shows that the LCT can be as low as 5°C for 70 kg animals under favourable conditions but 20°C for 40 kg animals in unfavourable ones. At the upper end of the thermoneutral zone there is a higher critical temperature, above which the animals are again uncomfortable and will expend energy in trying to keep cool. In fact, mist controllers (section 4.4) are employed as air coolers in piggeries. For meat animals the HCT lies in the 30°C region. The breeding sow, too, has a preferred range of temperatures (around 15°C) and young stock have their own, very critical requirements.

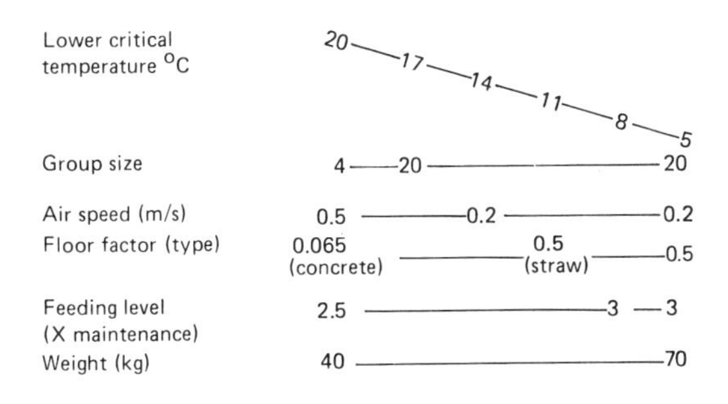

Figure 6.4 Lower critical temperature of finishing pigs: the effects of changes in five variables (Silsoe Research Institute).

Clearly, selection of the best temperature regime for each class of stock at each stage of development is a matter for calculation and/or sound judgement, with substantial penalties for error.

Once the desired temperature has been decided, however, it is obviously important to achieve it, and to maintain it under all conditions, if possible. In fattening piggeries, the step-control, automatic vent system of Fig. 6.1(a) was able to maintain the temperature to within $\pm 0.5°C$ at night but it was found that daytime variations of up to $\pm 1.5°C$ occurred due to the activities of stock and stockmen. This indicates the difficulty of maintaining close control in practice. However, in a well stocked building it proved capable of maintaining a 2.5°C *lift* above external subzero temperatures, without the addition of supplementary heat, and in hot weather it was able to restrict the lift to within 3°C at maximum ventilation. These findings pointed to the positive value of well sealed and insulated buildings, combined with efficient circulation of the internal air and close control of the ventilation rate. Nevertheless, as in the poultry sphere, there is a range of other fan controllers for the same purpose. Similarly, the ACNV unit already referred to can be used in naturally ventilated kennels for weaned pigs, to control the opening of the lid, as required in warmer weather.

Again, repeating points made in relation to poultry houses, temperature sensors should be inspected regularly for signs of deterioration since their unimpaired performance is crucial to the operation of the control system. There is also much merit in checking the performance of environmental controllers from time to time, using portable thermometers and air flow meters. For example, the minimum ventilation rates achieved by variable speed fans may not always match their specification in practice. In the pig sector RH monitoring is less critical, since the rule appears to be that high humidities are desirable, as a means to inhibit respiratory disease. However, if RH sensors are employed they too should be inspected occasionally for signs of corrosion or contamination.

The recommendations for farrowing sows are for a steady temperature of about 18 or 19°C, which is within the capability of a well designed and controlled system, but piglets need the additional warmth derived from a creep. This is vital to their survival immediately after birth, when they need a temperature of about 35°C, coupled with a draught-free ventilation system. The ventilation requirement remains for at least 8 weeks after birth but their ambient temperature needs to be reduced progressively to about 26°C at weaning and 20°C at 8 weeks. An infra-red heating system in the creeps dries the new-born piglets quickly and then provides an even heat where it is needed to warm the piglets and keep the flooring dry. This is achieved without overheating the sow to the detriment of her milk output. A control system such as that shown in Fig. 6.3 can be used to regulate the output of a radiant heater and it allows the producer to reset the required temperature to progressively lower levels as the piglets grow. The same sensor is employed, suspended where it will provide a representative value of the temperature in the piglet zone (but out of reach of these inquisitive animals). As before, the system can be used to control a row of creeps.

Ruminant environment

Ruminants are not as sensitive to changes in air temperature as non-ruminants and are best kept in naturally ventilated environments. However, where houses have insufficient ventilation, fan ventilation is employed. This is usually of the variable-speed (manually controlled) type, but time-proportional control (mentioned in relation to poultry housing) can provide the required purging action.

Alarm and paging systems

The UK livestock welfare regulations (1990) require automatic ventilation systems to be fail-safe and to be provided with alarms. In particular, mains failure must be accompanied by the provision of adequate ventilation. This can be achieved, *inter alia*, by the installation of ventilation panels which are held in the closed position by electrical actuators and which are released when the latter are deactivated, as already described. Alarm units with battery back up are available to warn of power failure, excessive temperature and other emergency conditions. In addition to operating bells or sirens, some automatically dial preset telephone numbers, in priority sequence.

Transport environment

Little has been known in detail until recently about the atmospheric, noise and vibration conditions to which animals are subjected during road transportation. Basic work is still in progress. However, experimental studies with poultry showed that the risk of thermal stress was greatest for birds located immediately behind the headboards of a lorry or trailer, during winter when the vehicles were covered. On that basis an on-board monitoring system has been developed, which predicts the conditions within those zones from sensors fixed to the vehicle itself. For commercial operation an algorithm was developed to predict conditions within the transport container, in order to relay warning signals to the driver. The same on-board data should provide the basis for improved control of ventilation on these transporters.

Vibration studies on these vehicles are establishing the frequencies to which birds are demonstrably averse, in order to establish criteria similar to those which govern human welfare regulations.

6.3 Animal identification

Automatic identification of animals follows here because many of the operations covered in later sections of this chapter are now being linked to the identity and life history of individual animals.

Automatic identification began with the larger dairy herds. Many different systems were invented and evaluated in the 1960s but eventually the transponder system proved the most practical and has been adopted internationally. The circuit carried by each animal is unpowered and is inactive until it is energised by a nearby coil which radiates energy at a low RF frequency. Once energised it radiates its own high-frequency signal, carrying the animal's binary code number, which is implanted in the circuit. That signal is received by an antenna, which relays it to a decoding circuit. The energising coil must be at the point where the animal has to be identified but the decoding circuit can be at a central point, where it can gather data from an array of antennae.

The system shown in Fig. 6.5(a) was one of those developed in the late 1970s, to be capable of uniquely identifying cows in herds of over 250 animals, if required. Standard data format for serial transmission (section 2.2) was employed, thereby taking advantage of existing hardware. In order to cope with large herds a 2-byte (16-bit) code was required. The formats of the code and transmission frame are shown in Fig. 6.5(b). The energising coil was driven by a 58 kHz oscillator with a power output of 1 W. The transponder required 4 mW to energise it, and that was collected by an internal ferrite rod (64 mm long, 10 mm diameter) with a 290 turn coil wound round it. The required amount of energy could be gathered at ranges

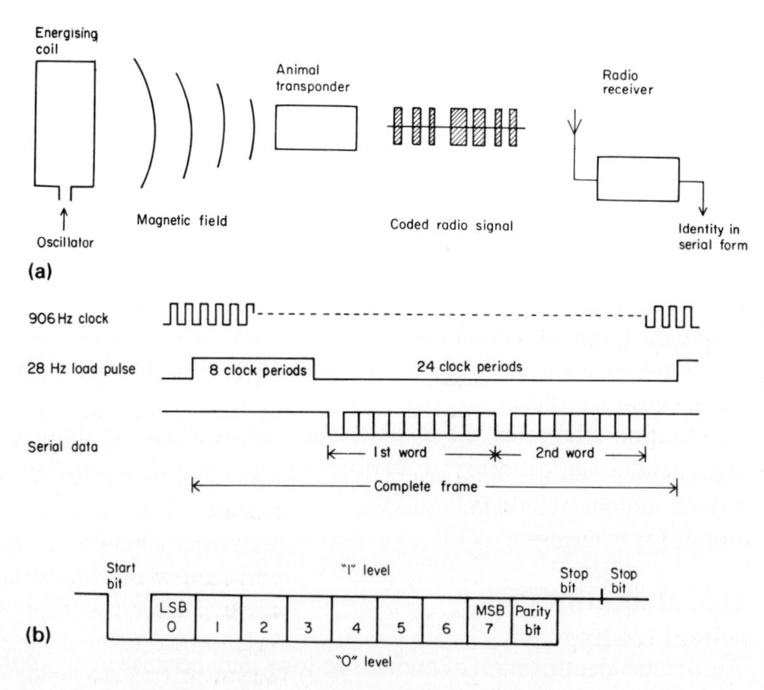

Figure 6.5 Animal transponders: (a) radio frequency identification system; (b) code generation and code format (Silsoe Research Institute).

up to about 1 m. The 58 kHz oscillations were also subdivided in the transponder to produce the 906 Hz clock pulses and 28 Hz frame loading pulses of Fig. 6.5(b). The frame was transmitted at an approved frequency of 26.995 MHz at a power of 0.5 mW, and with a bandwidth of 10 kHz. Although this carrier was modulated by rectangular pulses, as shown in Fig. 6.5(a), the transmission stayed within authorised limits for signal harmonics.

Most subsequent transponder systems have been variations on the radio frequency identification (RFID) concept, although some now employ magnetically encoded devices rather than the electrically encoded system typified by Fig. 6.5. In addition, collar-borne transponders have been giving way to ear-tag versions, starting with pigs and with breeding sows in particular. Implantable transponders have been under development since the early 1980s and have been used as implants behind the ears of pigs (where they are better protected) and in the rumen of cattle. Hand-held energiser/decoder units have been developed for quick identification of animals in groups, even when they are moving around. Reading takes only about 1 s. The number of bits in their codes rose from 16 to 64 and now they are available in 128-bit versions, which is more than adequate for a worldwide animal coding system.

The impetus for animal coding comes from the perceived need to trace the origins and life history of animals – especially meat animals – from birth to slaughter, in the interests of quality and security. The European Community's regulations will require dual ear tagging of animals, with details of their registered country of origin, herd number and individual animal number. Transponders can add to this basic requirement by facilitating the collation of data on individual animals, as well as acting as the animal's *pass key* to its feed ration or, very recently, to an automatic milking stall.

The subject of standardisation was taken up by ISO's Working Group WG3, Technical Committee TC23, in 1991. Subcommittee SC19 was set up to define an RFID standard. In 1994 the first draft of the standard on code structure was published. This was a precursor of ISO 11784 (1996). That structure is shown in Table 6.1. It will be seen that it refers to a 64-bit code, which is still sufficient for identification of animals worldwide. It also makes reference to ISO 3166 (1993) which defines standard country codes. The reserved bits (2–15) are intended for information that legislative bodies might need for additional information, while bit 16 allows the use of industrial transponders with more than 64 bits, which supply the potential to store information about an animal other than its permanent coding. A companion standard, ISO 11785 (1996) deals with the data format for transmission of 64-bit identifications.

6.4 Animal feeding

Most of the analysis of feed quality is done in the laboratory, where many measurements are performed by NIR spectrometers on both grain and forages

Table 6.1 ISO 11784: code structure.

Bit no.	Information	Combinations
1	Flag for animal (1) or non-animal (0) application	2
2–15	Reserved code	163384
16	Flag for additional data block (1) or no additional data block (0)	2
17–26	ISO 3166 numeric country code	1024
27–64	National identification code	274.877.906.944

(natural and ensiled). On farm, apart from the use of temperature and RH probes in stored feedstuffs, much of the instrumentation is concerned with weighing, directly or indirectly.

Cattle

For bulk feeding of cattle, front loaders can be fitted with a load sensor which registers the pressure required to lift the load. In one form the sensor is a pressure transducer. Its reading is taken automatically when the lift arm passes a fixed magnet, which operates a switch. An in-cab monitor produces an audible signal when that has happened. The reading can be added to a total or subtotal in one of several channels allocated to specific ration components or different loader attachments. Complete diet mixer/feed wagons employ load cells in the same manner as those on the trailer shown in Fig. 3.20. The wagons carry a load measuring and display unit at a point that is visible to the tractor driver, who can control loading and unloading from the cab. For loading they can be preset to the target amounts of each component of the diet, in the loading sequence, and will give warning when the corresponding accumulated weights have been closely or fully attained. However, in this application the accuracy attained is not so much a function of the load cells' capabilities as the degree of control attainable in loading the constituents.

Concentrate feeding of dairy cattle to their individual requirements (according to yield category or stage of lactation) was one of the first applications of RFID. As shown diagrammatically in Fig. 6.6, a cow looking for food at any time can put her head into the manger of an out-of-parlour feeder stall, where her transponder is activated by an energising coil below the manger. When her ID number has been recognised by a local or central controller, her programmed daily ration is looked up by the controller and if she is due for part of her day's allotment the controller actuates the dispenser above the manger. The allocated amount is dispensed volumetrically or by weight. The controller maintains a log of each cow's visit to the feeder and the amount dispensed on each occasion, which can be displayed on a VDU and printed out, as required. The log alerts the dairy

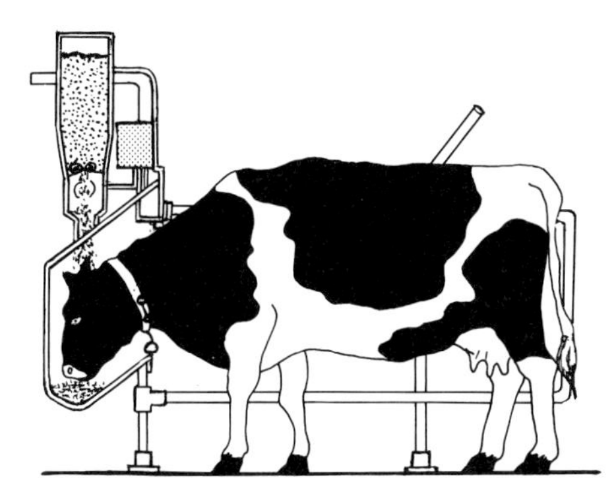

Figure 6.6 Automatic out-of-parlour feeder (Alfa Laval).

manager to *alarm* cows that are not taking up their concentrate ration, either because their transponder is faulty or because they have lost appetite for some reason. Cows may be allocated up to nearly 20 kg/day in this way, but possibly not more than a few kg at a time, spaced through the day. The dispenser should be calibrated from time to time, especially when there is a change of feed. The system was originally developed for feeding daily concentrates in the dairy parlour at milking but it could slow throughput. Where concentrates are fed in the parlour cows tend to be fed only as a fraction of their daily ration.

Calves have their own form of automatic feed control. Their liquid feed is made up from milk replacer, with included antibodies and water, possibly with the addition of cow's milk. One commercial controller, working with ear-tagged calves, can feed them a programmed mix of the above ingredients at the required temperature and concentration, using its own supply of water from a header tank. It rations the amount that each calf drinks daily and records that information, together with the calf's drinking speed. Since hygiene is a vital consideration it also has a wash cycle facility.

Pigs

Feeding costs in pig production far outweigh the cost of other inputs, therefore controlled feeding of least-cost rations is economically vital in this sector. The variety of feedstuffs fed to pigs and the variety of feeding systems employed are both very diverse but, in relation to the application of electronics, the field can be narrowed to two main types of installation, namely pipeline feeding in large fattening units and individual sow feeding. The former application is illustrated by Fig. 6.7, which shows the elements of a computer-controlled liquid feeding system. Liquid feeding is something of a misnomer for concentrated meal/water mixes in proportions of 1:2 to 1:2.5, respectively, but these mixtures can be pumped at rates of about 0.75 m/s through pipeline systems. They have the advantage for the farmer that pigs take in more food when it is in a wet condition and that pumpable mixtures with controlled nutritional content can be made up from many relatively low-cost ingredients. These may include skim-milk, whey,

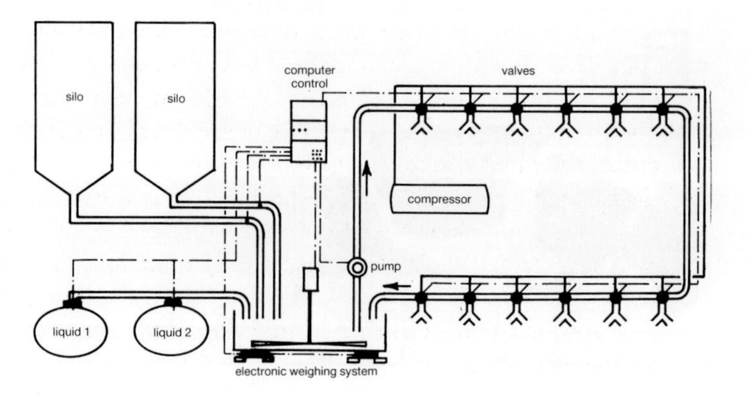

Figure 6.7 Pipeline feeding of fattening pigs: wet feed (Big Dutchman).

potatoes and moist grain. As shown in Fig. 6.7, water and other liquids, held in the tanks, can be blended with solids from silos in a mixing tank supported on load cells. In this way a required blend of pumpable consistency, and in the amount required, can be made up in the mixing tank. It can then be pumped round one or more pipeline circuits, fitted with a dispensing valve at each pig pen. The valves can be controlled electrically by the central computer, at low voltage (e.g. 24 V).

A large system such as this can blend up to 12 ingredients in three different solid:liquid concentrations and in amounts from 3000 l to 30 000 l, depending on tank size. The mixture can be pumped through up to eight circuits, and may control up to 500 valves. A feed can be dispensed by opening each of these valves in turn, or through a group of them, to feed a particular mixture to one set of pens. The rate of flow of the feed into the pipeline is derived from the change in weight of the mixing tank and the amount to be fed at each pen determines the time for which each valve is open. In a static installation it is possible to employ load cells with accuracies of $\pm 0.1\%$ of fro, which would make it possible, in principle, for a 1 t cell to measure a 1 t load to ± 1 kg. However, the actual accuracy achieved by the system will depend on the structural design of the installation, the environmental conditions and the effect of the mixing process on the load sensors. Ingredients included in smaller quantities will always be metered to lower accuracy, as in feed/mixer wagons, because the full-scale error (± 1 kg in the example just cited) will apply anywhere in the load cell's range. However, given the avoidance of extremes in the proportions of the mixture, together with a uniform mix and the dispensing of amounts close to tank capacity, metering and proportioning accuracies of within 2% are attainable.

The computer controller has a keyboard for entry of the percentage proportions of the feed ingredients and the amount of feed to be dispensed per pen. A change in the latter quantity with time can also be programmed into the computer, to provide a ration which increases automatically as the pigs grow in size. The computer then calculates the total amount of the mixture required at each feed (in kg) and the amounts of the individual ingredients needed to compound that feed. It also supervises the selection of the pipeline, where several circuits are installed, and the opening of the valves in the required sequences. The consumption of feed per valve and per circuit are stored and can be displayed. All the electrical functions of the system and the flow of material are monitored, to warn of any malfunction. Manual control is available if the automatic system should fail. This computer can be augmented with a printer or a VDU terminal in the farm office, through which supervisory programming and monitoring can be carried out. The terminal can also be used to compute optimal feeds and to maintain a sow calendar, where sows are being fed in this way, as well as performing other off-line tasks.

Smaller versions of the system do not contain a weigher but mix whey, water or skim-milk in preset solid:liquid ratios, employing an auger feeder to dispense the meal into the liquid. The controlled amount of feed pumped to each pen is monitored and stored in microcomputer memory, for display when required. Feed flow is monitored digitally, from the positive-displacement pump employed in these systems or by a separate flow meter.

Pipeline feeding of dry meal and pellets to pigs does not require comparable control systems. Instrumentation therefore has a more limited role, equivalent to that in poultry feeding, to be covered later. Individual feeding of group-housed dry sows, using the equivalent of the dairy cow's out-of-parlour feeder was introduced in the 1980s, primarily as a welfare measure – allowing the sows room for exercise and natural behaviour, but avoiding aggressive competition for food. It also provided the opportunity to control individual feeding, which would be difficult to achieve otherwise in a loose housing environment. However, this still requires more elaborate measures than are needed in the dairy environment, to make the feeding stall bully-proof.

Poultry

Electronic monitoring and control for dry feeding of housed poultry employs methods applicable to pig feeding as mentioned above, and to cattle feeding, for that matter. Commonly, the feed is supplied from bulk hoppers with legs supported on load cells. Loads up to 30 t may be involved. Load cells can be fully exposed to the weather, therefore their design and construction must be of high quality. Figure 6.8 shows two strain-gauged cells for this application, one in its mountings for support of a hopper leg. The horizontal slots in the free cell signify that it embodies a shear-sensing technique which renders it insensitive to side loads, by a 200:1 factor. It is also tolerant of 300% fro overload, although half of that figure could also alter its calibration. Its total error from non-linearity, hysteresis, creep and total random effects is about ±0.1%, and its temperature compensated range is –10 to +60°C (operational range –40 to +80°C). Thermal effects total ±0.01%/°C. Environmental protection is IP66/67 standard (i.e. highly water proofed). The energising four-lead cable is screened, with

Figure 6.8 Load cells for silo weighing (Thames Side Scientific).

connection to earth at the meter end only, to avoid interfering earth loop signals. High tensile pins connect the cell to the cast mounts.

Calibration of the complete load cell assembly can be done by adding a known mass of feed to the hopper or by check weighing small batches of feed unloaded from the hopper, using a relatively low-range weigher of high accuracy.

Inside the poultry house, feed from the main hopper is distributed via pipelines to feed-line hoppers and then to the birds. In the main hopper and at feed-transfer points (pipeline-to-pipeline, or pipeline-to-feed-line hopper) capacitance type proximity sensors can be sited as high- or low-level switches, to control the drive motors in each line. Examples of their application are shown in Fig. 6.9(a). These units, sealed to IP 67 standard, have a sensitive face which detects the presence of the fed material. The capacity-sensitive circuit controls an internal relay, which enables the unit to start or stop an external motor when it is actuated, as required in a particular application. Some also have an adjustable time delay.

Another means for control of the rate of feed dispensed is provided by the weigher shown in Fig. 6.9(b). It has a dust- and water-proof enclosure with an upper loading point and a lower discharge point, which makes it suitable for mounting between a feed supply auger and a poultry feed conveyor. An internal drum is suspended from two load cells (one at either end) which provide the input to a microcomputer-based controller, mounted on the base of the enclosure. This can be programmed to stop the overhead auger when the drum has received a

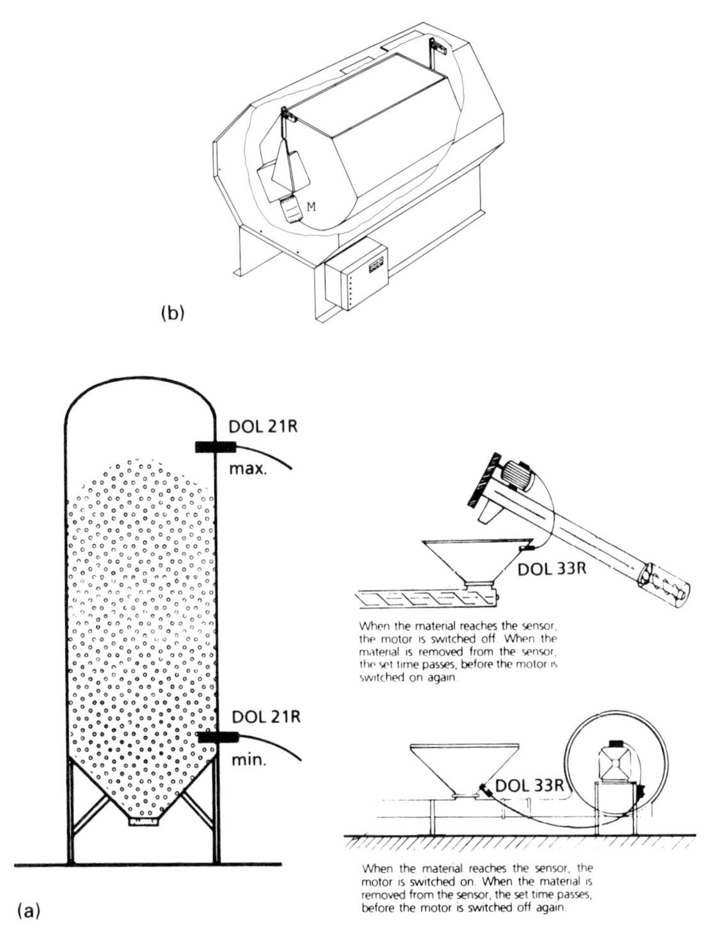

Figure 6.9 Poultry feeding: (a) capacitance level sensors; (b) batching weigher (Dolphine Ventilation).

preset amount of food, in the range 10–30 kg. It then starts a motor (M in the diagram) which revolves the drum and discharges its load. When the drum has been emptied its tare weight is automatically measured and the cycle will begin again unless a preset limit on the amount distributed has been reached. The control unit displays the cumulative amount fed out since it was last reset and it closes a contact for each 10 kg delivered. The latter facility provides information for a remote counter or computer.

Controlled feeding at rates up to 4 t/h can be achieved and with a timer input restrictive feeding up to about 10 kg is possible. Failure of the feed supply results in an alarm signal, whereas if no supply has been required for about 5 minutes the drum is automatically inverted. The latter operation makes the system ready for cleaning without risk of water or debris entering the drum. A cap at the top of the drum allows the user to insert four 5 kg masses for calibration purposes.

A dual-feed version of this weigher is also available, for combined grain and concentrate feeding. The amounts of the two components dispensed are recorded separately.

Feed-rate monitoring can be augmented by electrical monitoring of the birds' rate of water intake. This requires a positive-displacement water meter which produces a digital pulse at the end of each cycle. The total food and water intake can then be compared with the total output, in terms of liveweight gain, through regular bird weighing.

Sheep

Research work has shown that up to 70% of lamb deaths are influenced by inadequate feeding in the ewes' late pregnancies. Their requirement can rise to three times that of the maintenance level during early lactation. The consequent relevance of pregnancy detection to ewe feeding is raised in section 6.5.

6.5 Animal monitoring

Regular monitoring of animals is necessary for checks on their productive capacity and for their welfare. In the former context much of this routine requires surveillance by the stockman or woman, using established criteria such as *condition scoring*, together with herd records. In the latter context careful observation, coupled with experience, should detect problems at an early stage. Nevertheless, as indicated at the end of the preceding section, automatic measurement and control systems can provide valuable numerical data for decision making as well as reliable control of routine operations. In addition, round the clock surveillance of livestock is rarely practicable and the time available for individual attention diminishes as herd sizes increase. Therefore automatic surveillance is an area of increasing importance.

Animal weighing

Regular weighing of meat animals has clear benefits for producers wishing to monitor feed/meat conversion rates throughout the animals' growth, as well as their weight at market. Unfortunately, weighing of animals individually is time consuming and it can be hazardous when they are both large and lively. The operation requires a well planned arrangement of races and pens if it is to proceed smoothly, and even then there can be a problem with the occasional difficult animal. All of these features are deterrents to regular weighing. In addition, recording itself is not always an easy task. The traditional spring balance is capable of adequate accuracy (± 2 or 3%), if it is of good design and well maintained, but the movement of an animal in a weighcrate often makes it difficult to read. Heavy damping can be incorporated, but this makes the balance sluggish and can introduce errors unless the damping system is also well maintained.

Electronic weighing cannot overcome the problems of handling a stream of animals into and out of a conventional weighcrate. However, it can speed up the reading and recording of weights. The essential feature of the electronic livestock weigher is that it can average the output of a load cell or cells supporting the weighplatform and then present the result as a steady reading with an accuracy of about ± 1 or 2%, in most cases. Averaging can be done by an analogue or digital circuit over 2 or 3 s, but the display is usually digital. As in all livestock weighing, the accuracy obtained in practice depends upon the care with which the weighcrate is set up. It must be on firm level ground; all pivot points and linkages must be clean and free to move and the movement of the weighplatform itself must be unobstructed. When the load is measured by an electrical transducer, the usual care must be taken with cables, connectors and the quality of the power supply. Equally, the transducer itself needs to be protected against environmental extremes, which can degrade its performance. Periodic checking of the system with known weights is also essential.

The digital meters supplied with these weighers have auto-zero, taring and memories which store the number of animals weighed in a session, together with their total and average weights. Serial outputs are available for printers or PCs.

Weighing of dairy cows is less frequent, although dairy farmers recognise that the weight trend of their animals can help to reveal health and nutrition problems before they become too serious. The problem of handling these animals is less difficult, too, by virtue of the RFID transponders now in wide use. Trials have shown that it is not difficult to weigh them automatically as they leave the milking parlour, by installing a walk-through weigher in the exit passage. Here they are normally walking steadily without crowding, as seen in Fig. 6.10(a), while the load pattern on the platform is recorded. Sample recordings are shown in Fig. 6.10(b). From left to right, a cow rubbed its neck against the weigher's frame as it passed through; the cow passed through cleanly, making a rapid exit; the cow was hurried through by a man. In general, there is a well defined peak when the whole of the cow is on the platform, preceded by a lesser peak as she places her front feet on it, and a variable (to non-existent) peak as she leaves it. The central peak is correlated with her code number (read as she leaves the weigher) and added to the time history of her weight in the parlour computer. If, for any reason, an invalid weight is registered by the load sensing system the computer records a blank reading. It is also programmed to reject readings that are apparently valid but depart by a preset amount from a running average of the cow's weight. A virtue of weighing after milking is that it reduces the degree to which recordings are affected by diurnal fluctuations in the cows' body weights. These can amount to 15 kg or more, largely as a result of changes in gut fill, apart from the milking process, therefore it helps to take measurements at regular times, such as post-milking.

The weigher shown in Fig. 6.10(a) is a standard weighcrate with a suspended

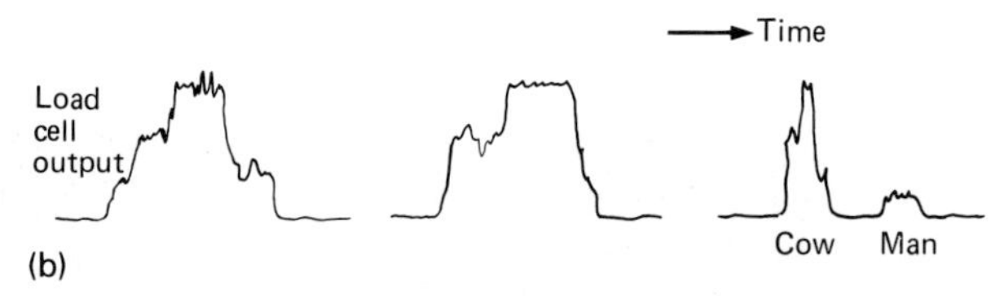

Figure 6.10 Unconstrained livestock weighing: (a) walk-through weigher for dairy cows (Fullwood and Bland); (b) load cell output patterns.

weigh platform of about 2.5 m length. That ensures a recording of a second or more, which is sufficient for weighing to an accuracy of about $\pm 2\%$ in most cases. A longer platform would allow a longer averaging time, which could improve that accuracy (if required) but interference by a following cow would be more likely. The load sensor employed with the above weigher is a tension-link, strain-gauge load cell with 0.5 t rated capacity, which occupies the place where there would have been a spring balance in the original crate. The load cell is environmentally protected against exposure to wind and sun and it communicates by overhead cable to the parlour computer.

The need for regular poultry weighing was referred to in the context of poultry feeding, but manual weighing of a sample of the birds in a broiler house – which may hold 20 000 birds – is time-consuming and stressful for both the birds and the operators. These considerations led to the development of automatic poultry weighers in the 1980s. Their purpose was to collect a representative sample of the birds' weights over the period of their growth, in order to provide the information that the producer needs to achieve weight targets by specific dates. That required an array of specially designed perches, distributed around the building at a

(c)

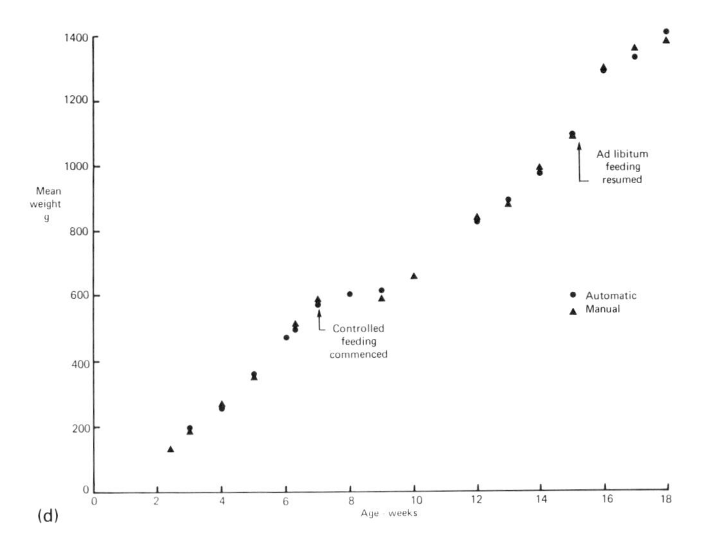

(d)

Figure 6.10 (cont.) (c) weighing perch; (d) automatic vs manual weighing, pullets (Silsoe Research Institute).

convenient height above floor level, as in Fig. 6.10(c). The perches shown in the figure were suspended from a strain-gauged weighbeam, via an elastic element which protected the beam against shock loads. Lateral and rotational constraints were applied but the perch was allowed some movement in those directions to discourage individual birds from taking up long-term residence on it. In order to accommodate the considerable change in the size of the birds during their development, three perches of different widths but the same mass were employed. As with the automatic weigher for dairy cows, the attendant computer discarded suspect readings (e.g. two birds on the platform or competing for it) by applying an *out of bounds* algorithm. A graph obtained from one batch of birds is shown in Fig. 6.10(d).

The new development of the 1990s has been the application of machine vision to weighing animals by indirect means. This subject is dealt with later in the section.

Oestrus and pregnancy

The economics of livestock production depend on conception rates and dates and on birth rates, to a considerable degree. Synchronised oestrus, artificial insemination and embryo transplants have introduced major control measures, while monitoring methods include mounting detectors for oestrus and progesterone tests for pregnancy. Still, the use of electronic pedometers for automatic detection of the onset of oestrus in dairy herds has been adopted in larger herds. These devices employ a motion sensor, such as a mercury switch, which is attached to one of each cow's legs above the hock. The number of movements detected by the device is stored by an internal microprocessor, to be read by an interrogating device at intervals, or fed directly into the parlour computer at milking time. Peaks in activity occur over about half a day at oestrus and are typically about two to three times the baseline activity calculated on past data. Although this is not a certain method for detecting the onset of oestrus (figures of about 80% have been quoted) it helps to reduce the number of missed *heats*, which in time shortens the herd's calving index (or interval). In particular, the pedometers may help to pick out cows with a very short oestrus period (often high yielders) with a tendency to show symptoms during the night. Some of these devices also incorporate transponders which provide the cow's identification.

Monitoring for oestrus by measurement of the electrical conductivity of a cow's vaginal mucus was developed in the 1970s, and cervical probes are manufactured for that purpose. These can show the optimum time for insemination, but of course the approximate time of oestrus needs to be estimated first.

Following insemination, ultrasonic pregnancy detectors have been employed for many years on cattle, pigs, sheep and goats, in order to confirm pregnancy or to provide an early indication of the need to return an animal to service. They are in the form of a high frequency (MHz) ultrasonic transmitter/receiver, linked by a cable to a meter with a display and audible response, as in Fig. 6.11, or to a

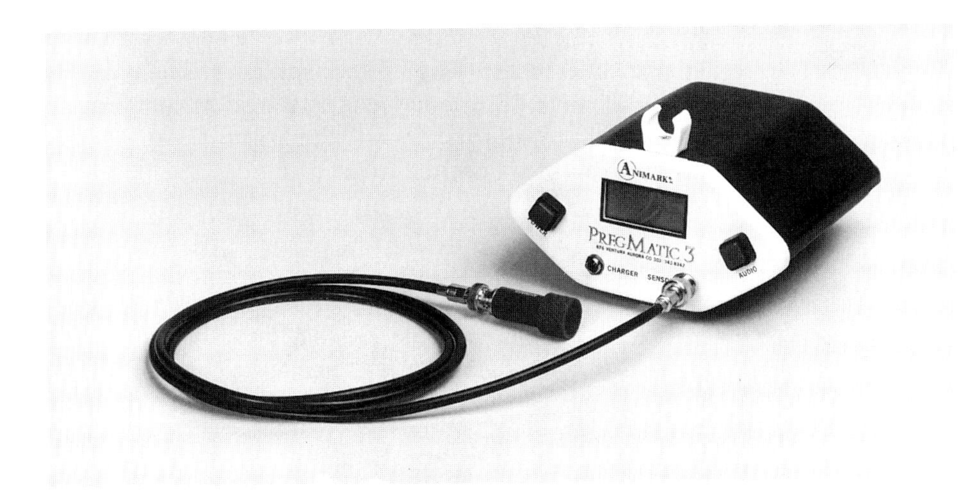

Figure 6.11 Pregnancy detector (Animark).

headphone set which provides an audible signal. The probe must make good contact with the animal's skin and there must be a good acoustic match at the interface, to avoid the reflection of a large proportion of the ultrasonic energy at that point. Bland greases or oils are applied to the skin for that purpose. The monitoring site is the animal's abdominal region and the detection method is now generally measurement of the Doppler frequency shift (section 1.10) caused by foetal heart beats or pulsations in the blood flowing through the animal's uterine artery. Skill is needed in using these instruments, although signal processing can remove unwanted *noise* from the echoes. Nevertheless, the measurement is quick and simple in principle, and its pregnancy detection rate can be close to 100%. It is even more reliable for detection of non-pregnancy.

With sows, pregnancy can be detected within 4 weeks from service, thereby enabling a return to service by the next heat, if required. Ewes can be diagnosed in 70 days, while multiple pregnancies can be detected and monitored throughout the pregnancy, so enabling the producer to tailor the ewes' feeding levels to the number of fetuses that they are carrying (see section 6.4). That can reduce mortality rates while using feed economically. Monitoring of multiple fetuses is not always easy, however, and in consequence mobile monitoring stations have been used to scan each ewe with an array of transmitters/receivers, to produce an image of the fetuses (as in human obstetrics). The ewe has to be placed on her back in a cradle for this purpose. With cows, an external probe can detect pregnancy at 5 months, but this can be reduced to 7 weeks by a rectal probe.

At parturition a pessary with a built-in temperature sensor is available to give warning when a cow is giving birth, via a radio link with a range over $\frac{1}{3}$ km. This can be valuable in reducing calf mortality, by ensuring that farm staff are at hand in the event of complications.

A more recent development is the use of ultrasonic imaging for the detection of the sex of embryo calves, about 8 weeks after conception, when their sexual organs are beginning to differentiate. This information can influence the farmer's planning in a number of ways. Ultrasonic imaging is also used to assist in the recovery of oocytes from the ovaries of cows with fertility problems. After culture of the oocytes, embryos can be transferred to the cow.

Animal health and welfare

Here instrumentation and control systems have featured more extensively in milk production than in any other section of the livestock industry. Studies of automatic methods for oestrus determination have been matched by those on mastitis detection over the same time periods, by virtue of the suffering of cows with clinical mastitis, as well as the serious economic loss for which the disease is responsible. Pressure on producers to combat mastitis by routine hygiene, dry cow therapy, maintenance of milking machines and culling of animals that are subject to repeated infection at clinical level has been increased by legislation. In Europe, EU Directive 92/46 makes milk with a somatic cell count (SSC) greater than 400 000/ml unsaleable, since that level is a sure indication that bacteria causing the disease are present.

Mastitis detection

Simple and effective filtering devices are available for detection of clinical mastitis, which results in clots in the milk after 2 or 3 days. However, early detection and treatment of the condition, based on milk conductivity measurements, has been shown to reduce the ensuing SSC by as much as 50% compared with treatment following the appearance of clots. Measurement of the electrical conductivity of milk as a means to detect mastitis infection originated with the work of Linzell and Peaker in the early 1970s. The problem has been that there is no simple relationship between the two quantities. Conductivity measures ionic composition, as stated in Chapter 1 and elsewhere in this book, i.e., it is determined by the level of salts in the milk. That level is undoubtedly related to bacterial infection of the cow but – to repeat earlier statements again – the conductivity of liquids is very temperature sensitive. Temperature changes accompany oestrus and metabolic disorders such as ketosis, as well as udder infection. Therefore both parameters must be measured.

As in many other areas, the ability of the computer to analyse complex data online has made it possible to add routine milk conductivity measurements to the monitoring facilities offered by the parlour computer. Initially, manufacturers have incorporated conductivity electrodes in their milk meters. This has the advantage that at the monitoring point the milk can be free of entrained air, which would render the measurement unreliable. Some of these meters also measure the conductivity of the milk from all four quarters of the cow's udder, independently, over the milking period. That is an important facility since severe mastitis infection is frequently limited to one quarter, while the cell count in the other quarters is at subclinical level. In fact, algorithms based on combined measurements of quarter milk conductivity (QMC), milk temperature and milk yield may provide the means to monitor for oestrus and the severity of mastitis simultaneously, and automatically.

The current method of data processing for mastitis is similar to that for oestrus detection. Baseline data for each quarter are established by tracking conductivity to establish a cow's individual pattern. Then a threshold is set to trigger an alarm when the conductivity rises above that level. Confirmation may be obtained by a simultaneous drop in recorded milk yield.

Teat cleaning

Teat cleanliness and disinfection is among the hygiene measures that can be taken to reduce the incidence of mastitis. Pre-milking checks on the cleanliness of teats before the application of teat cups is a necessary operation, accompanied by cleaning and drying of the teats if required. This is a legal requirement under the above mentioned EU Directive. It can be complemented by post-milking disinfection. Automatic teat spraying as the cow leaves the parlour is a relatively straightforward operation, as exemplified in Fig. 6.12. An array of floor-mounted jets (S_1–S_3) is positioned to give all-round coverage of the teats. One directs its spray vertically upwards and the other two aim at the udder from offset directions. The jets are activated in stages by the four light-beam/photocell assemblies (L_1F_1 to L_4F_4) in order to avoid indiscriminate and costly use of disinfectant. Initially, the passage of the cow breaks all four light beams, as shown in Fig. 6.12. Then, when F_1 is reilluminated, followed by F_2 and F_3, S_1, S_2 and S_3 are energised in turn. When F_4 is reilluminated the sequence is reset and all three sprays are off.

Regular washing and drying of teats before milking is not a general practice, since this can cause teat sores which harbour infection as well as forming a focus of transferable infection at teat ends. That presents a problem for automatic milking (section 6.6) since some automatic means must be found to decide

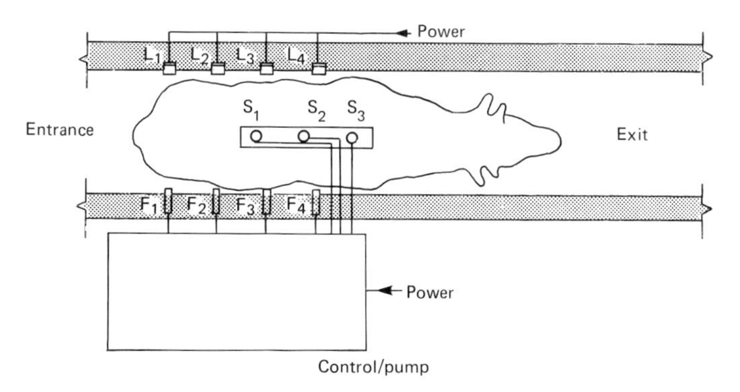

Figure 6.12 Automatic teat spraying (Gascoigne).

whether teats require cleaning or not. Suitable methods of imaging and image analysis are being sought.

Lameness

This is another major problem in dairy herds, which has led to research originating in the 1980s. By the early 1990s researchers using multi-element pressure plates and computer imaging systems were able to show the complex pressure patterns set up in a cow's foot when she stands on a hard surface. This is still an area of research activity, but it may provide the specification for a foot pressure monitoring system in the dairy parlour, to alert the dairy worker to cows needing inspection.

Surveillance

The video camera, in different sizes and price ranges, now provides round the clock surveillance of animal enclosures, when linked to a central VDU display and a recording system. Several cameras can be shared by one VDU in some commercial systems. Image analysis, based on CCD cameras, is another area of development, aiming to identify animal behaviour patterns which could provide early warning of environmental problems, as well as conditions such as oestrus in individual cows. The techniques can be exemplified by Fig. 6.13. This is a thresholded image of four pigs whose outlines have been determined by an edge-detecting algorithm. From these shapes the computer software can learn the shape of pigs and, from another algorithm, estimate their weights by determining prescribed widths and lengths in their images. Reference to this technique was made earlier in the section, in relation to pig weighing. Apart from that the technique can show whether the pigs in a pen are active, spread out or huddled together. If they are huddled closely, that may mean that they are below their lower critical temperature (see section 6.2).

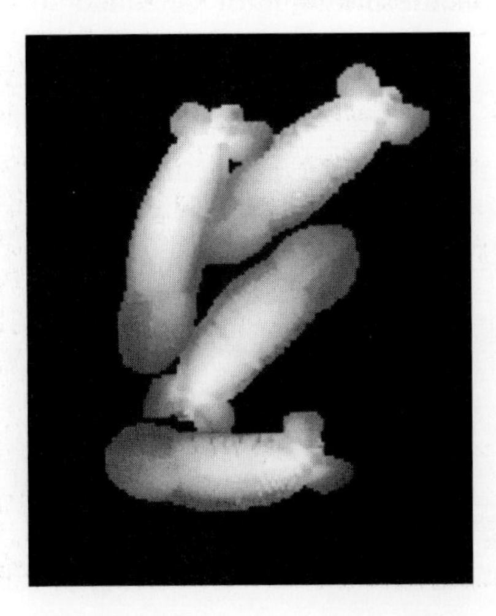

Figure 6.13 Machine vision: animal monitoring. Thresholded image of pigs (Silsoe Research Institute).

6.6 Cattle milking

Milking machine control

Precise control of the vacuum applied to milking clusters is vital to machine milking and to the welfare of the cows, as already noted. The high vacuum level, the pulsation rate and the pulsation ratio (Fig. 6.14(a)) are all important factors, which can be checked by test meters with vacuum transducers. Vacuum levels are held at about half atmospheric pressure, as indicated in Fig. 6.14(a), and pulsation rates at 50 to 60/s, independent of ambient temperature. The pulsation ratio is generally set to 50% at 50 pulsations/s and 70% for 60 pulsation/s. These settings are based on a compromise between higher milk flow and risk of damage

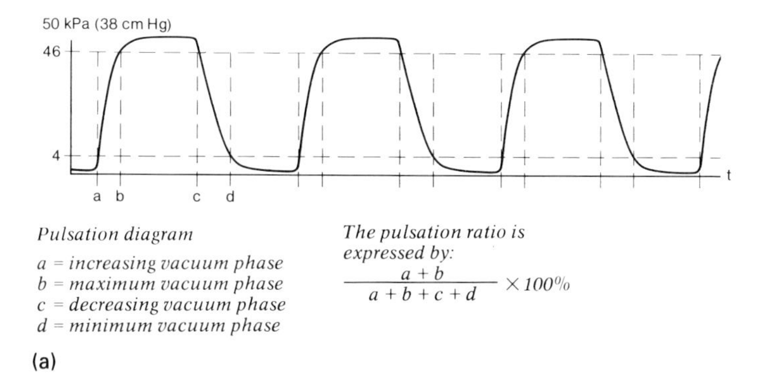

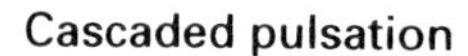

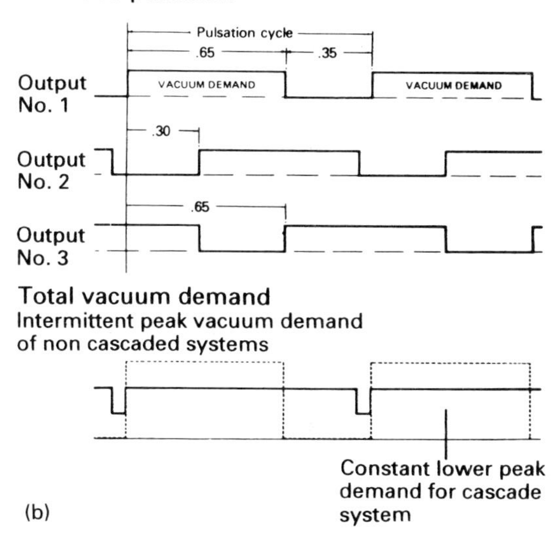

Figure 6.14 Milking machines: (a) pulsation diagram (Alfa Laval); (b) cascaded pulsation (Gascoigne).

to the cow's teats. Milking can begin at the slower pulsation rate (stimulation phase) before moving to 60 pulsations and maximum milk flow. The ISO is in the process of establishing standards for these and other aspects of milking installations.

The vacuum system can maintain better control if the call on it varies as little as possible. To that end the pulsation cycle at each cluster can be synchronised in pairs, although this can make the pulsation ratio in the two udder halves slightly different. The imbalance, expressed as a percentage of one pulsation, must be kept below 5%. Another form of load spreading is employed where a number of pulsation generators are distributed at milking points. Cascaded pulsation generators create three staggered outputs, which even out overall vacuum demand, as shown in Fig. 6.14(b).

Milk meters

Automatic measurement of each cow's milk yield is based on traditional recording jars, shown schematically in Fig. 6.15(a), or direct-to-line meters which convey milk from the milking cluster to the pipeline, for transfer to the bulk tanks. All meters must conform to the standard set by the International Committee for Recording the Productivity of Milk Animals (ICRPMA). For acceptance in specified tests the mean value of the meter's indications must be within 2% of a reference yield (or 200 g for yields less than 10 kg). Its coefficient of variation for reproducibility must be no greater than 2.5% of the reference yield (or its standard deviation must be no greater than 250 g if the reference yield is less than 10 kg). In addition to these rules the meter must be easy to clean during pipeline decontamination with hot detergent and disinfectants. The pipeline meter must also be able to cope with milk flow rates up to about 10 kg/min, while an ISO standard limits the pressure drop that it produces in the line. For example, this must be not more than 3 kPa at a flow rate of 3 kg/min.

The metering method employed in Fig. 6.15(a) is to weigh the jar and its contents with a strain-gauged support beam. The drag of the attached flexible pipes represents a disturbance to this weighing system but careful disposition of the pipes minimises the effect. ICRPMA standards can be satisfied, with the additional benefit that the reading is directly in mass, rather than volume measurement, which has to accommodate milk density variations. One design of direct line-to-line meter also incorporates a load cell which supports an internal weighing cup. The cup collects, then releases 300 g batches of the milk, which are totalled by the associated meter. On release, a representative sample of each batch is diverted into a chamber, from which it can be collected in a sample bottle, for quality analysis. The meter drains completely after each cow's milking, to provide the total mass of her yield. Calibration of the weighing system is achieved by applying a known weight loading via a screw-off cap at the side.

Volumetric sensing to meet the ICRPMA standards of accuracy is found on both recording jars and direct-to-line meters. One type of recording jar sensing is

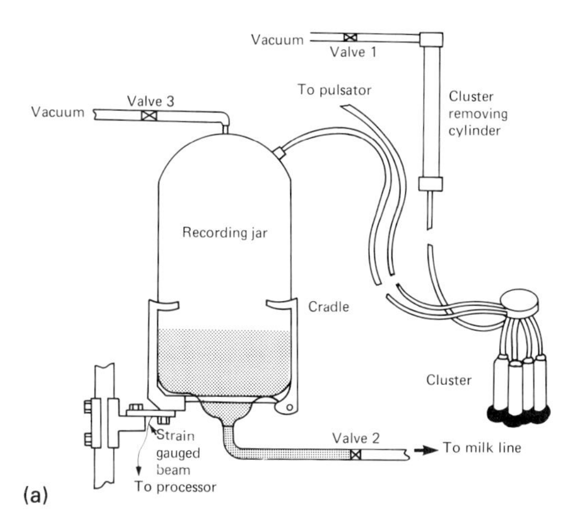

Figure 6.15 Milk recording: (a) recording jar with strain-gauged support beam (Silsoe Research Institute); (b) recording jars with float level sensors (Nedap Poeisz); (c) direct to line meter, with sampling bottle (Fullwood).

shown in Fig. 6.15(b). The sensor in each jar comprises a vertical, central, water-filled rod with an internal ball bearing and an external, cylindrical float which contains magnetic material. These move in unison as the float tracks the level of milk in the jar. The level of the ball bearing is measured internally through ultrasonic ranging by a transmitter/receiver in the head of the probe. The base of a recording jar can also be fitted with a pneumatically operated piston, which is driven from side to side in a cylindrical chamber, to deliver a 100 g batch of milk into the pipeline at each stroke. That quantity is the increment recommended by ICRPMA. Direct-to-line meters also have a metering chamber which fills and discharges repeatedly. That can be achieved by fitting the chamber with three electrodes at different levels. Milk then enters the chamber until there is conductance between the central and uppermost electrodes, at which point the inflow is cut off and an outlet valve is opened. When the conductance path between the central and lowest electrode is broken the outlet is closed and the inlet reopened. Sampling facilities are provided, as in the corresponding direct weighing meter (see Fig. 6.15(c)).

The information on milk flow rate provided by all of these meters can be used to determine an end point for a cow's milking and to operate automatic cluster removal (ACR). The unit shown in Fig. 6.15(a) serves to illustrate the sequence. Milk recording by the parlour computer starts when the operator presses a button at the milking place. The tare weight of the empty jar is recorded by the computer. Valve 1 closes to release the cluster for attachment to the cow; valve 2 closes and valve 3 opens. The accumulating milk yield is recorded until, near the end of milking, its rate of change falls below a preset limit (say 0.25 kg/min) chosen to avoid over – or under – milking of the cow. After a delay (say 20 s) the end of milking sequence is initiated. The cow's yield is stored by the computer and the valve positions are reversed. The cluster is then removed automatically by the vacuum cylinder.

Bulk tanks can be fitted with electronic sensors, too. A milk tank manufacturer provides a level sensing probe with an external float, similar to the unit used in recording jars, to measure the volume stored. This has been designed to provide a check on the amount of milk delivered to a collecting tanker, as determined by the tanker's turbine flow meter (the basis for payment to the producer). In the UK the former Milk Marketing Board set the target accuracy for the turbine meter at $\pm 0.35\%$. A figure of $\pm 0.2\%$ is claimed for the probe.

The parlour computer

Repeated reference has been made to the parlour computer in this section. In fact, monitoring and control in the dairy parlour have become part of a SCADA system (section 2.2). Programmable parlour units such as that shown in Fig. 6.16 gather data from the *field-level* devices (transponders, pedometers, milk meters) and control vacuum pulsation ratios, ACRs and concentrate dispensers. They also communicate with a central herd management computer which maintains individual cow records and downloads instructions on any animals that require special attention. This computer works out feed rations for individual cows on the basis of their yields and stage of lactation. It can also provide data on treatment for a variety of ailments. The above parlour unit's keypad enables an operator to call up information from the central computer, using the 18 code

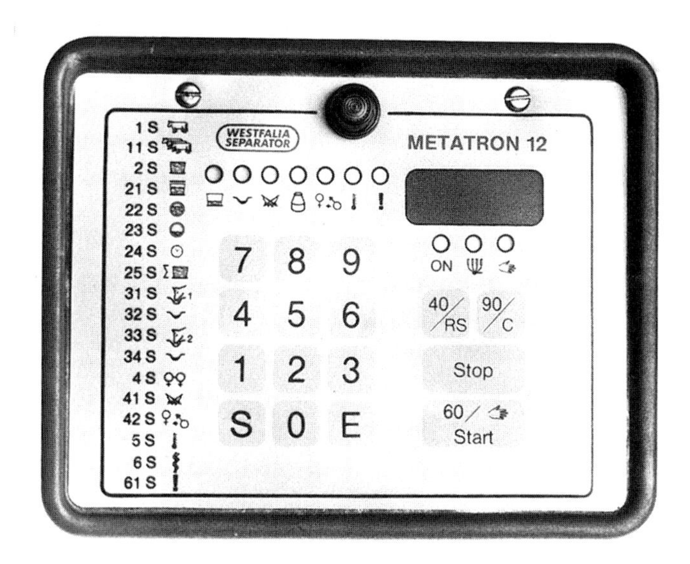

Figure 6.16 Dairy parlour monitoring and control unit (Westfalia).

numbers listed on the left of its face. The seven LEDS above the keypad draw attention to actions required when a particular cow is identified (e.g., 'note milk yield' or 'do not milk'). The large, central LED above them flashes to draw attention to the fact that one or more of them is lit. It also flashes slowly when a cow's milk flow has reached the threshold set for the ending of her milking, and it remains lit at the end of the milking, to show the operator that she is ready to leave the parlour. The remaining display provides four-digit numerical data (meter readings, etc.). A conductivity alert has been added to the original specification of this unit.

Automatic milking

This has been developed to farm evaluation stage over the past decade. Its origins lay in physiological studies which showed that twice a day milking was optimal neither for cows (calves naturally suckle five or six times per day) nor milk output. Milking three or four times daily is not practicable on most farms. Technology offered prospects for fully automatic milking on demand, equivalent to concentrate feeding on demand. Work began on the first – crucial – task, namely automatic attachment of teat cups. There are as many ways of doing this as there are automatic milking systems (AMS). All have a robot arm which applies teat cups individually, either by ultrasonic or optical sensing (including a camera in the latter category). Some assist the robot's search by building up data on each cow's teat conformation over successive milkings, to provide a computer model for steering the arm to the point of which the arm's sensor can complete the docking manoeuvre. Others use an arm to make a new search at each milking. The information that it gathers enables the computer to steer a second, cup-bearing arm to the point at which the cups can be attached. Individual removal of teat cups is a feature of some systems.

Then the stance of the cow is important, to allow the arms clear space for scanning and attachment of the cups. All systems apply lateral and longitudinal constraints to the cow's movement. Some have a raised floor at the front of the stall or cubicle since the cow's posture then keeps the teats clear of her legs for much of the time. Another system causes her to stand with her rear legs splayed. Most position the robot at the side of the cow but one inserts the arm from the rear between her back legs, on the grounds that it is less disturbing to her. Maximum avoidance of injury to the cow underlies the design of a pneumatic system which picks up and applies one cup at a time. The use of pneumatics makes the arm light and fast moving, and it withdraws immediately if kicked, then returns automatically.

Cattle receive some concentrate feed on entering the stall and are identified automatically at that time. Once they are in position some systems brush or sponge their udders automatically, too, although selective cleaning has advantages, as stated in section 6.5.

Farm experience has been that cows can soon become accustomed to an AMS and will visit it voluntarily several times a day. However, not all have teats that are easy to connect to a teat cup. Furthermore the AMS must fit into a planned layout, including feeding and lying areas. In fact, cow traffic control is incorporated in one experimental system. Individual cows are allocated an allowed frequency of visits to the AMS. This requires them to gain access to the milking unit only after being accepted at an identification barrier. Those cows who are not allowed through are diverted back to the feeding and lying areas.

6.7 Livestock product testing

Here two techniques for quality grading of animal products can be cited, without straying into the much larger sphere of food processing.

Ultrasonic testing

The ultrasonic pulse-echo technique described in this and other chapters was first used to determine the depth of back fat on live pigs in the 1960s. Although the acoustic differences between fat and lean (muscle) meat are fairly small they are significant enough to produce detectable echoes at fat/lean interfaces when ultrasound is transmitted through the body of an animal. Similarly, as would be expected, a strong echo is generated at a tissue/bone boundary. The principle of the measurement is shown in Fig. 6.17(a), which represents a cross-section of the back of an animal and the path of a narrow beam of ultrasound, derived from a transmitter/receiver operating somewhere in the 1 to 10 MHz range. The diagram

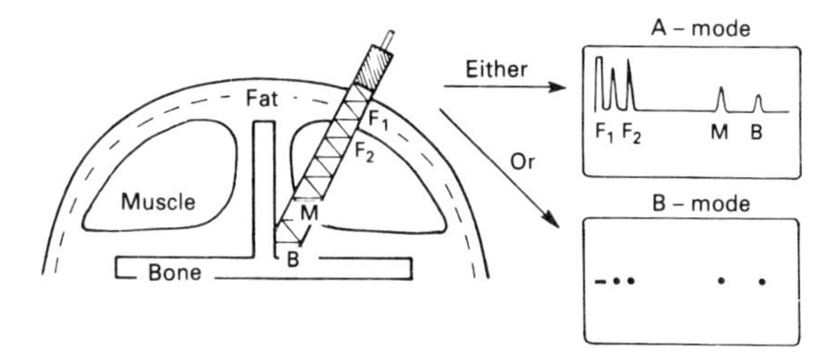

F_1 = interface between 1st and 2nd layers of fat
F_2 = interface between 2nd layer and muscle
M = ventral boundary of muscle
(a) B = bone

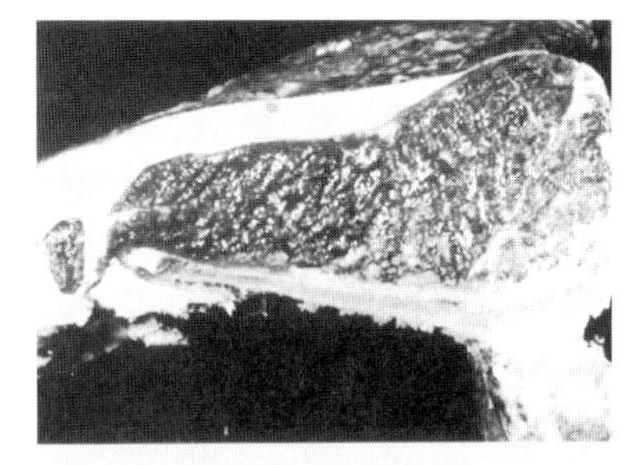

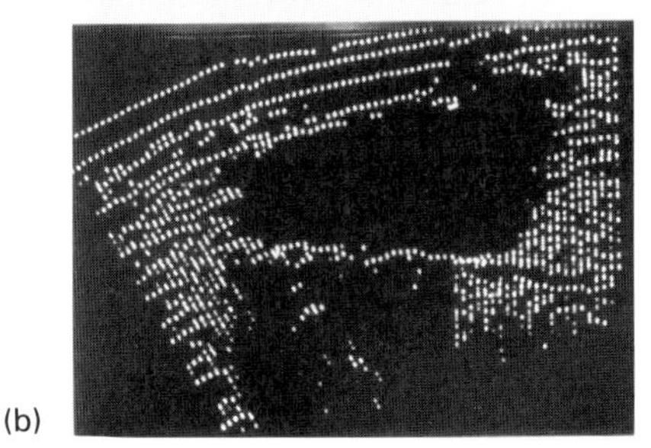

(b)

Figure 6.17 Ultrasonic testing of meat animals: (a) pulse-echo; (b) B-scan and comparison with carcase.

also shows the two common ways of presenting the echo pattern on the face of a cathode ray tube (CRT) or standard VDU. The 'A' mode shows the transmitted pulse (on the left of the trace) and the successive echoes at boundaries F_1, F_2, M and B. Note that the back fat contains two layers, with an acoustically distinct region between them – hence the F_1 echo. The return signals from boundaries M and B are inevitably smaller, although the acoustic mismatch at them is large, because, through accumulated attenuation, only a fraction of the radiation returns to the transmitter/receiver. The transmitter/receiver is coupled to the animal's skin with bland oil, as for pregnancy detection. There is a reflection at the skin/fat interface, too, but this is too close to the transmitted pulse to be distinguishable. The A scan shown in Fig. 6.17(a) is based on a time scale (i.e., the time taken for echoes to return) and not a distance scale. The distance travelled

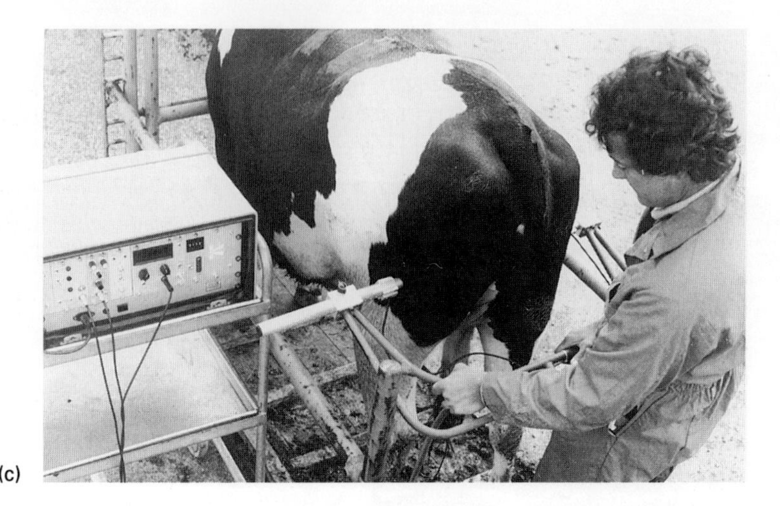

(c)

Figure 6.17 (cont.) Ultrasonic testing of meat animals: (c) transmission method (Meat Research Institute, Bristol).

by the ultrasound in a given time is dependent on the speed of sound in the animal's tissue and this varies with the acoustic properties of the tissue. For example, at 37°C the velocity of acoustic waves in fat and muscle is about 1.45 and 1.6 km/s, respectively. Therefore, errors in estimation of the thickness of the fat and lean tissues can be of the order of 10%, unless speed corrections are made. In the B mode the return signals modulate the brightness of the scanning electron beam in the CRT, therefore the echoes are displayed as bright spots on the screen. The latter mode is the more versatile because it can be used with scanning transmitter/receivers to build up a two-dimensional picture of the fat/lean distribution within the animal (Fig. 6.17(b)).

The use of this equipment requires practice and some skill but, taken with regular weighing, it can provide the pig producer with valuable information on the condition of finishing pigs. With that information the producer can make more accurate judgements of the market readiness of the animals. The method has also been used with success on fattening lambs – and without damaging the fleece by shaving the animal at the points of contact with the transmitter/receiver. This has been demonstrated by workers in New Zealand, using a 5 mm diameter transmitter/receiver instead of the more usual 25 mm unit. The small 5 MHz probe can be coupled to the animal's skin effectively with oil or petroleum jelly. More precise measurement of fat depth is required than with pigs, because the fat layers may be only a few mm in depth. At these ranges, the uniformity of the skin thickness is important, too. In addition, movements due to the nervous reactions of the animals make the pulse-echo pattern very unsteady. In consequence, the measurement is far more difficult to perform manually. However, if the probe is coupled to a microcomputer the latter can process the data and produce a steady display. With the aid of computer processing, back fat can be measured to the nearest 0.5 mm. Inevitably, the equipment is more costly than the standard units used with pigs.

This technique is less appropriate for cattle, since the disposition of fat and lean tissue is far less clear-cut than in pigs or sheep. Consequently, interpretation of pulse-echo patterns is again very difficult. Therefore the method of ultrasonic transmission has been developed for beef animals. It was shown by work at the former UK Meat Research Institute that the speed of ultrasound passing through mixtures of fat and lean tissue is inversely related to the proportion of fat, i.e. the more fat, the slower the speed. The most suitable measuring technique is to determine the transmission time of ultrasonic pulses from one side of the animal's hind limbs to the other, as shown in Fig. 6.17(c). The sensor is in the form of a G clamp, which is adjusted to fit the animal being tested. The width of the gap is measured automatically by an optical coding device on the adjustable rod which carries the ultrasonic receiver. The receiver assembly is on the left of the animal in Fig. 6.17(c), together with the main electronic unit. The latter energises the ultrasonic transmitter on the far side of the animal, then computes its fatness from the transmit-receive time and the width of the gap – the time increasing with fatness. Again, the equipment is costly relative to the pulse-echo units used with pigs, but it provides a rapid, simple measurement which can be employed by beef producers to monitor their selection programmes, as pig producers have been able to do for a long time.

Egg candling

Normally, optical inspection of the interior of agricultural produce is difficult, in the context of commercial grading, because of its opacity. One exception is egg candling, which can be performed manually on a continuously moving conveyor. The eggs are illuminated from beneath the conveyor and they are almost perfect diffusers, glowing with the colour of the light. The use of light at or near a wavelength of 575 nm (yellow light) is optimal for detection of one of the prime causes for rejection of eggs, i.e., blood spots or blood distributed in the albumen. Blood has an absorption band in that region of the spectrum, and therefore the amount of light transmitted will be visibly affected by its presence in the egg. In practice, sodium lamps can be used since they radiate light at about 590 nm.

The outline of part of a fully automatic egg grading and packing installation is shown in Fig. 6.18. Under computer control the system can take eggs directly from the laying houses or from trays and automatically feed them on to a conveyor which takes the eggs to the candling booth (on the right of the diagram). The rate of feed of the eggs is controlled to keep the conveyor filled. In the booth the under-conveyor sodium lighting is concentrated by reflectors on to the underside of the eggs, which are held semivertically, in rows, between diabolo-shaped spacers. By this means the intense light is not beamed directly into the operators' eyes. The operators' task is to identify reject eggs and to separate them from the remainder. Hitherto this has required them to pick up the rejects and

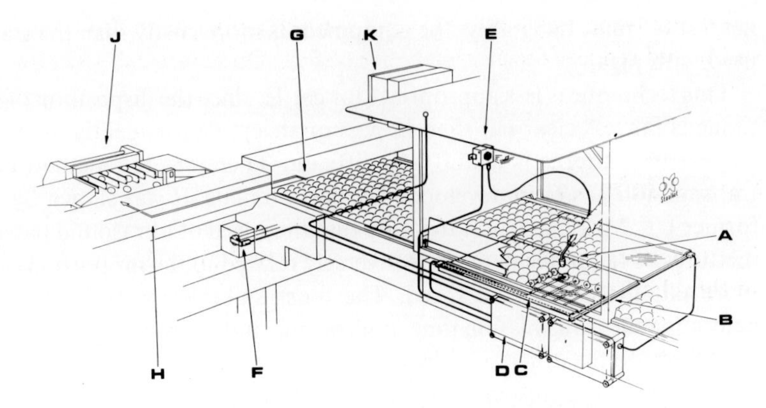

Figure 6.18 Egg-candling: semi-automatic system (Staalkat).

place them in an adjacent tray. However, it is also possible to semi-automate the operation, by a method originally developed at the former Scottish Institute of Agricultural Engineering for use in potato grading. The operators hold a device, sometimes known as a *wand*, with which they tap the rejects as they pass by. This can be seen at A in Fig. 6.18. Tapping an egg generates a pulse from a piezo-electric transducer disc in the head of the wand and this is communicated by the wand's cable to the main electronic circuit D. Then D feeds back a radio-frequency oscillation to a transmitter coil in the wand. Between the sodium lights and the conveyor a cross matrix of electrical coils (B, C) is mounted, covering the whole width and length of the presentation area. One set of coils runs parallel to the separate lanes of the conveyor; the other runs laterally and has the same spacing as the rows of eggs. When the wand radiates its RF signal an output voltage is generated in the coils in its vicinity. The maximum outputs come from the lane and row coils that lie under the egg which has just been tapped. At the same time, the operator is given an audio acknowledgement from unit E in the booth, which confirms that the main unit has registered and memorised the location of a reject egg. Unit D also receives information on the movement of the conveyor from the shaft monitor F, as in the potato graders described earlier. Therefore it can operate the rejection mechanism in the appropriate lane when the row containing the reject egg reaches the delivery point. Rejects take path H to J where they are packed in trays. The remaining eggs are channelled into their appropriate weight grades, determined at G. The main computer displays (and can print out) the number and percentage of rejects per consignment, together with the total and average weight/grade (to 0.1 g) and other management information, or the system can be linked to a larger, supervisory computer.

The benefit of the wand has been to increase the capacity and quality of manual grading considerably, because the operators no longer have the distraction of removing eggs by hand and they can concentrate on the task of identifying the rejects. Also, they are spared the direct view of the sodium light to which they are subjected when eggs are removed from the conveyor in the booth. This is fatiguing and reduces the efficiency of grading. Operators can also perform the task seated.

6.8 Further information

The following published papers and reviews broadly follow the sequence of topics in this chapter.

Section 6.2

Bruce, J.M. & Clark, J.J. (1979) Models of heat production and critical temperature for growing pigs. *Animal Production*, **28**, 353–69.

Randall, J.M. (1993) Environmental parameters necessary to define comfort for pigs, cattle and sheep in livestock transporters. *Animal Production*, **57**, 299–307.

Randall, J.M., Streader, W.V. & Meehan, A.M. (1993) Vibration on poultry transporters. *British Poultry Science*, **34**, 635–42.

Section 6.3

Geers, R. (1994) Electronic monitoring of farm animals: a review of research and development requirements and expected benefits. *Computers and Electronics in Agriculture*, **10**, 1–9.

Street, M.J. (1979) A pulse-code modulation system for automatic identification. *Journal of Agricultural Engineering Research*, **24**, 249–58.

Section 6.4

Puckett, H.B., Hyde, G.M., Olver, E.F. & Harshbarger, K.E. (1973) An automated individual feeding system for dairy cows. *Journal of Agricultural Engineering Research*, **18**, 301–307.

Section 6.5

Animal weighing

Filby, D.E., Turner, M.J.B. & Street, M.J. (1979) A walk-through weigher for dairy cows. *Journal of Agricultural Engineering Research*, **24**, 67–78.

Turner, M.J.B. & Smith, R.A. (1974) Electronic aids in fatstock weighing. *Journal of Agricultural Engineering Research*, **19**, 299–311.

Turner, M.J.B., Gurney, P. & Belyavin, C.G. (1983) Automatic weighing of layer-replacement pullets housed on litter or in cages. *British Poultry Science*, **24**, 33–45.

Oestrus

Maatje, K. & Rossing, W. (1976) Detecting oestrus by measuring milk temperatures of dairy cows during milking. *Livestock Production Science*, **3**, 85–9.

Marshall, R., Scott, N.R., Barba, M. & Foote, R.H. (1979) Electrical conductivity probes for detection of estrus in cattle. *Transactions of the American Society of Agricultural Engineers*, **22**, 1145–51.

Animal health and welfare

Distl, O. & Mair, A. (1993) Computerized analysis of pedobarometric forces in cattle at the ground surface/floor. *Computers and Electronics in Agriculture*, **8**, 237–50.

Fernando, R.S., Rindsig, R.B. & Spahr, S.L. (1980) Electrical conductivity of milk for detecting mastitis. *Illinois Research*, **22**(1), 12–13.

Linzell, J.L., Peaker, M. & Rowell, J.G. (1974) Electrical conductivity of milk for detecting sub-clinical mastitis in cows. *Journal of Agricultural Science*, **83**, 309–25.

General

Frost, A.R. (ed) (1997) Special issue: livestock monitoring. *Computers and Electronics in Agriculture*, **17**, 139–261.

Section 6.6

Bull, C., Mottram, T. & Wheeler, H. (1995) Optical teat inspection for automatic milking systems. *Computers and Electronics in Agriculture*, **12**, 121–30.

Ordolff, D. (ed) Special issue: robotic milking. *Computers and Electronics in Agriculture*, **17**, 1–137.

Section 6.7

Beach, A.D., Tuck, D.L. & Twizell, R.J. (1983) Ultrasonic equipment for measurement of backfat on unshorn live sheep. *Ultrasonics*, **21**, 184–7.

Johnson, E.K., Hiner, R., Allsmeyer, R.H., Campbell, L.E., Platt, W.T. & Webb, J.C. (1964) Ultrasonic pulse-echo measurement of livestock physical condition. *Transactions of the American Society of Agricultural Engineers*, **7**, 246–9.

Miles, C.A., Fursey, G.A.J. & Pomeroy, R.W. (1983) Ultrasonic evaluation of cattle. *Animal Production*, **36**, 363–70.

Chapter 7
Ergonomics and Safety

7.1 Introduction

Health and safety at work are the subjects of national and international codes and regulations of many kinds, which are being applied ever more rigorously. Within the EC six Directives on health and safety at work were introduced in the early 1990s. These dealt with the six areas shown in Table 7.1. In the UK, new Health and Safety at Work regulations came into force in 1993. These require all employers with five or more employees to have a written health and safety policy, in which risks are assessed and measures for dealing with them are set down. These measures include the necessary training and health monitoring of employees. In parallel, Provision and Use of Work Equipment regulations require employers to use only equipment that is suitable for both the task and the environment in which it is used. Further, the equipment must be properly protected and maintained. Specific reference is made to control systems and control devices. This involves the IP classification of instrumentation protection (section 2.4).

Table 7.1 EC directives on health and safety at work (HSE).

• Health and safety management
• Work equipment safety
• Manual handling of loads
• Workplace conditions
• Personal protective equipment
• Display screen equipment

Safety considerations also include those related to farm products themselves. In the UK, as soon as an edible crop is harvested it is classified as food and becomes subject to food safety regulations. That has obvious implications for the crop spraying and storage processes described in previous chapters.

Unfortunately, agriculture and its related industries have had a poor safety record over many years. In particular, the UK's annual lists of fatalities on farms continue to catalogue 'operator error' as the cause in most cases. In this context, instrumentation and control systems can be a potential cause of accidents, as well as a means to reduce them by providing the operator and (employer) with information through which the risk of accidents can be reduced. With regard to safety hazards posed by such systems, references have been made in previous chapters to diagnostic and fail-safe features. In part these have been aimed at protection of the user, through early warnings of impending trouble or by prevention of operator actions that could precipitate an accident. In addition, there is a range of instrumentation specifically employed for safety monitoring. Examples are given in this chapter. Since most farm accidents involve machines and equipment, the chapter is initially concerned with monitoring of hardware; the remainder deals with the buildings environment more generally.

7.2 Machines and equipment

Vehicle stability

One of the most hazardous tasks for a tractor driver is working on sloping land, where a multiple overturn can cause serious injury or death even when the tractor is fitted with a safety cab. The conditions in which this can occur depend on many factors – the steepness of the slope, the ground cover and soil condition, the type of tractor and implement or trailer, trailer loading, the direction in which the machine is travelling relative to the slope and the operator's control actions. Work done at the former SIAE and continued in the Scottish Agricultural College (SAC) showed that a grass-covered slope is the most hazardous surface. On this surface a 4.14 t, 2 WD tractor could begin to slide with locked wheels down a 11° slope in the worst case. The corresponding figure for a 3.73 t 4 WD tractor (front wheel drive engaged) was 20°. The best case for the latter was 34°. Initial work produced a safe descent slope meter, shown diagrammatically in Fig. 7.1. The pendulum sensor (coupled to the potentiometer) was near-critically damped so that dynamically it acted as a decelerometer in braking tests. In slowly varying conditions its output indicated the upward or downward slope on which

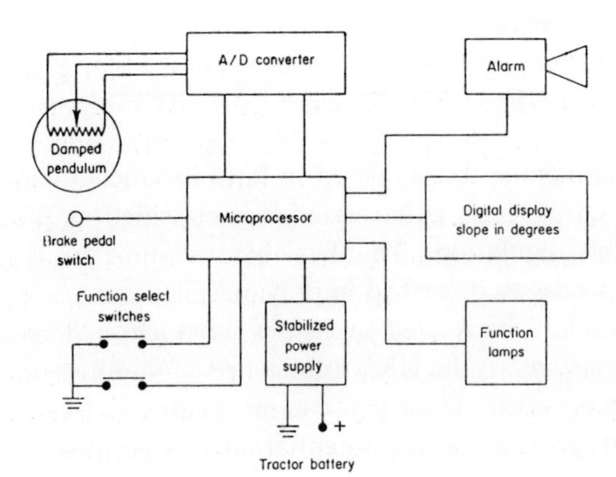

Figure 7.1 Safe descent slope meter (Scottish Institute of Agricultural Engineering).

the tractor was working at any time. In locked wheel braking tests on level ground the microprocessor recorded the maximum angular displacement (deceleration) of the pendulum. That displacement was shown to be the critical angle for hillside working on a similar surface, i.e., for work on hillside grass a prior braking test would be made on level grass. The microprocessor retained the critical slope in its memory and operated the alarm if the driver came close to working at that slope.

Subsequent work produced a stability index which could be incorporated in an on-board safety instrument such as this. In addition, a static stability measurement technique was developed which requires only a 10° slope, a single portable weighpad for measuring wheel load, an inclinometer to measure the slope and a hand-held calculator. The weighpad is placed under the same wheel when it is in the uphill and downhill positions (equivalent to using a 20° slope). The stability angle is then computed from a simple trigonometrical formula involving the two weights. These techniques also have application to forestry work in hilly terrain.

Noise and vibration

Longer term hazards to the operator include subjection to acoustic noise and low-frequency body vibration. It became clear in the 1950s and 1960s that tractor drivers were subjected to unacceptably high levels of both these hazards. Considerable reductions in their levels have been made possible since then, through the introduction of *quiet* cabs and seat suspensions in particular. At the same time more stringent regulations have been imposed. For example, in 1994 the EC began the development of Directives on *physical agents* at work, applying initially to noise, vibration and non-ionising electromagnetic radiation. Table 7.2 summarises (and so simplifies) information available on draft provisions for the first two agents, available in mid 1994. The broad aim of the proposed Directives is to specify a series of levels of exposure above which specified action is required.

Noise

Noise measurement and analysis is a complex process that requires trained operators. Nevertheless, a brief introduction may serve to put the figures in Table 7.2 in context. Figure 7.2(a) shows contours of perceived equal loudness of pure (sinusoidal) tones at different sound pressures over the audible frequency range. Perceived loudness is a subjective assessment, of course, but these contours provide a starting point for noise measurement. First, they show the necessity to present sound pressure on a logarithmic scale, since human hearing is decreasingly sensitive to change as sound pressure increases from the lower threshold of audibility to the upper threshold of painful loudness. These thresholds are conventionally taken as 20 μPa and 200 Pa, respectively, yet the perceived change in loudness is approximately the same for a tenfold increase in pressure anywhere in that range, for most audio frequencies, as can be seen by reference to Fig. 7.2(a).

There, a tenfold increase equates to 20 dB, through the relationship between sound pressure (p) and sound pressure level (SPL), written as L_p:

$$L_p = 10 \log \left(\frac{p^2}{p_0^2}\right) \text{dB}$$

where $p_0 = 20\ \mu\text{Pa}$. [Note The decibel, dB, is one tenth of the bel – a logarithmic power ratio.] This modifies to

$$L_p = 20 \log \frac{p}{p_0} \text{ dB}$$

Therefore, when $p = p_0$, $L_p = 0$, and when $p = 20$ Pa, i.e. $\frac{p}{p_0} = 10^6$, then $L_p = 120$ dB. The threshold of pain is not shown in Fig. 7.2(a), but at that point $L_p = 140$ dB.

The ear's response to tones of different frequencies shows that frequency analysis of noise is required for detailed studies of its composition and abatement. However, for general surveys the standard instrument is the sound level meter

Table 7.2 EC directives on daily exposure to (a) noise (b) whole-body vibration (1994).

(a)

Daily personal exposure	Peak sound pressure	Notes
60 dB(A)	—	Target upper limit for rest areas
75 dB (A)	—	Risks to be minimised. Workers to be informed of risks
80 dB(A)	112 Pa	Workers to be informed on protective and control measures, and supplied with personal protective equipment if they request it. They also have a right to a hearing check if they want it
85 dB(A)	112 Pa	Training for workers on protective and control measures. Noise assessments and, when necessary, measurements. Employers to establish a programme of control measures. A record of the programme is not expressly required but the need to provide the programme to safety representatives and workers suggests a record would be a practical necessity. Workers' representatives and workers to receive results of the exposure assessments and the programme of measures for control
90 dB(A)	200 Pa	Ear protectors must be used Systematic health surveillance by or under the responsibility of a doctor must be carried out Marking hazard areas
105 dB(A)	600 Pa	Activities to be declared to the 'authority responsible' which must take appropriate measures to control the risk

Table 7.2 *Continued.*

(b)

Daily (8 h) exposure, by reference to ISO 2631 (except where indicated)	Notes
0.25 m/s^2	Threshold level Risks to be reduced to the lowest achievable level Information for workers on risks
0.5 m/s^2 or 1.25 m/s^2 (1 hour average)	Employers to establish a programme of control measures A record of the programme is not expressly required but the need to provide the programme to safety representatives and workers suggests a record would be a practical necessity Information for workers on protective and control measures Training in implementation of measures Vibration exposure assessments and when necessary measurements Workers entitled to regular health surveillance Workers' representatives and workers to receive results of exposure assessments and the programme of measures for control
0.7 m/s^2	Systematic health surveillance must be carried out If vibration levels cannot be adequately reduced exposure to be controlled by restricting exposure time. For a period of 5 years from implementation member states may grant derogations from this requirement if the state of the art does not allow the value to be respected
1.25 m/s^2	Activities to be declared to the 'authority responsible' which must take appropriate measures to control the risk

(SLM) which incorporates the A weighting shown in Fig. 7.2(b). It can be seen that the frequency weighting curve is another approximation, intended to make the meter's response similar to that of the ear. Measurements taken with this weighting are recorded as dB(A). On this scale 60 dB(A) is equivalent to the level of normal room conversation, 75–85 dB(A) approximates to the level in a busy street, 90 dB(A) to the noise of a heavy truck at close range and 105 dB(A) to a similarly close pneumatic drill. Surveys of exposure of agricultural workers to noise were made for the UK Health and Safety Executive (HSE) in the late 1980s. These showed that 18% of working days were spent at levels above 85 dB(A) and nearly 9% at much higher levels.

The figures quoted in Table 7.2 are for daily personal exposure. Few people work in a constant noise environment, therefore the concept of the equivalent sound energy level over a working period has been standardised. Thus L_{eq8} is a commonly quoted figure, that is the total sound energy to which an operator has been exposed over an 8 h working shift, averaged over those hours. Other

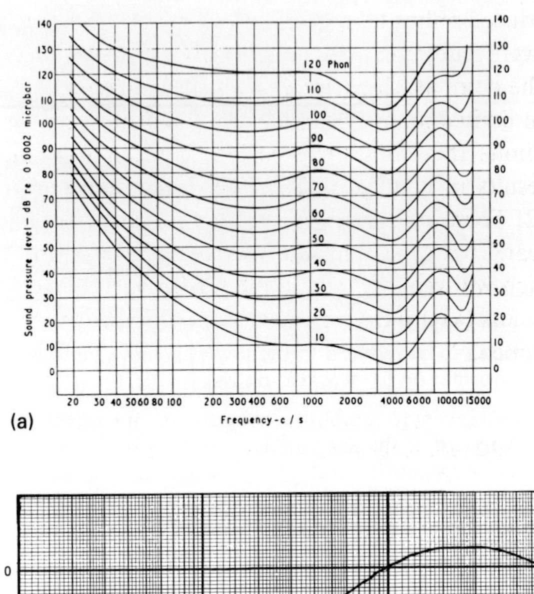

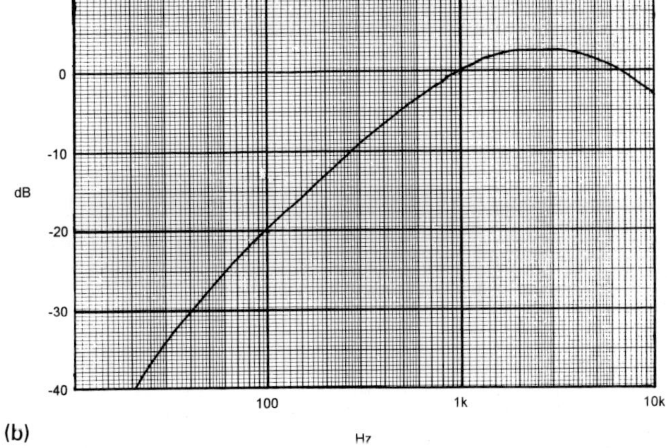

Figure 7.2 Noise measurement: (a) equal loudness contours; (b) A-weighting.

provisions are made when the noise is impulsive, and peak levels become more important than averages for hearing conservation. Hence the peak pressure figures in Table 7.2(a).

IEC standards 651 and 804 apply to SLMs for conventional and integrated measurements, respectively. Capacitance microphones are universal. The more advanced instruments, with microprocessors incorporated, provide L_{eq} measurements as well as SPLs. These are tools for the specialist user, but modestly priced SLMs for general assessments of the noise environment are available and some instrument hire companies supply IEC grade SLMs, with standard noise generators which provide overall calibration of these meters.

Vibration

Earlier in this section it was stated that suspension seats in tractors made improvements in vehicle ride possible. In fact, field surveys showed that these seats produced only a temporary reduction of the ride vibration to which tractor drivers were subjected. Increased driving speeds increased vibration levels beyond the point at which any seat suspension could be effective. The reason for this is the limited amount of seat movement that can be accommodated, which in turn limits the vibration isolation that can be achieved. Figure 7.3 shows survey results in relation to the ISO's recommended 8 h limits, mentioned in Table 7.2. These measurements were performed with accelerometers mounted on the seat. It is clear that many field operations are performed under conditions which will require 'systematic health surveillance' under current EC proposals, unless (a) drivers slow down or (b) improved tractor or cab suspensions are developed.

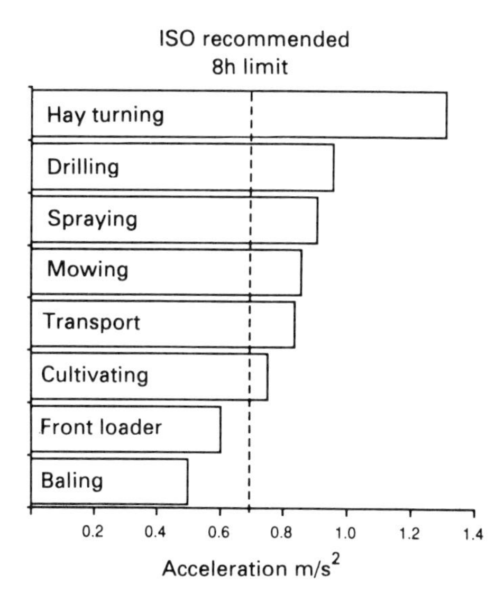

Figure 7.3 Ride vibration: ISO recommended 8-hour limit in relation to agricultural tasks (Silsoe Research Institute).

Displays and controls

As integrated monitoring and control on field machines develops on the lines described in Chapter 3 the ergonomics of the man–machine interface (MMI) becomes more important, as indicated in section 2.4. Displays need to be readily visible, clear and simple, with graphics and mimic diagrams rather than varying digits, where there is a choice. Control switches and keypads need to be big enough and near enough for easy reach and operation, without distracting the operator's attention from steering and observation of the field task for more than a second. Alarm signals should be specific as well as clear.

In this context, the possibility of control by voice commands seems unlikely in the tractor noise environment but recent research has shown that neural networks can be trained to recognise vocal signals in that environment.

Driverless machines

In the light of developments such as the autonomous vehicle described in section 4.2, the question of safety is bound to arise. In that connection, although there were no official regulations at the time (1980) an HSE advisory document set down provisional safety requirements for automatic tractors, which included the following:

- The operator can start, drive to site, operate, transfer to automatic control, override automatic control as necessary and stop the machine safely.
- Automatic operation does not create any hazards for the public including children.
- The automatic control system is neither hazardous in itself nor liable to receive conflicting signals, instructions or control from outside influence which would lead to a dangerous malfunction.
- The tractor shall be prominently marked on the front with diagonal black and yellow stripes in accordance with British Standard 5378: 1976, and should also carry a flashing amber light. There should be prominently displayed on the tractor a notice stating that it is automatically controlled.
- Any control for setting the tractor in motion shall be so designed and constructed that the tractor cannot be set in motion by an unauthorised person, e.g. by provision of a removable key. The control shall also be positioned so that the tractor cannot be set in motion by any person standing in the path of any part of the tractor, or of any of its equipment carried or moved over the ground.
- Contact detection equipment shall be provided so that if an object is contacted or detected, not less than 550 mm from the ground, the tractor should come to rest with the engine stopped before the most forward part of the tractor, or its attachments or load reach the point at which the object was positioned prior to the time of detection. The detection equipment shall surround the front and sides of the tractor and be as wide as any attached equipment. It shall extend at least to the vertical plane which passes through the centre of the tractor rear axles. An object will be regarded as having collided with the outfit when a force, or combination of forces greater than 40 N is detected.
- The tractor shall automatically be brought to rest with its engine stopped in the event of the receipt of a false signal or equipment failure.
- Manual or automatic control shall be designed so that selection of one method of control completely eliminates the function of the other. In addition the control systems shall be so designed that when the driver is sitting in the tractor seat, energisation of the automatic system shall not be possible.
- In the event of failure of the control equipment such that the tractor or any part of its attachments departs from its correct direction and is unable to determine the correct position, the tractor should be brought to rest with its engine stopped.

- Means should be provided on the exterior of the tractor for safely bringing the tractor to rest with the engine stopped.
- The working area or field should be enclosed by a fence or barrier and the tractor should not be used where any public right of way passes within these enclosures. A warning notice that an automatic tractor is in use should be displayed at all entrances to the working area.
- The tractor should comply with the requirements of the Health and Safety at Work Act 1974 and its relevant statutory provisions.

7.3 Buildings environment

Electrical equipment

The Institution of Electrical Engineers (IEE) is responsible for wiring regulations in the UK and it has recently produced a Code of Practice for in-service inspection and testing of electrical equipment in the workplace. The code was endorsed by the HSE and it provides information that enables employers to comply with the UK Safety at Work Act and the Electricity at Work regulations which meet EC requirements. These last mentioned regulations emphasise the requirement for regular inspection and testing of portable electrical appliances (IEE recommends 6-monthly intervals) and there is now a range of portable appliance testers (PATs) for that purpose. These cover insulation and earth continuity, in support of visual assessments of the condition of plugs, sockets, cables and other aspects of the integrity of the appliance. Some also test the residual current devices (RCDs) which back up circuit breakers or fuses, to give protection against electric shock or fire caused by earth leakage currents. Portable appliance testers are usually supplied with a stock of self-adhesive 'Pass' labels which can be used to indicate the last date of testing of an appliance. They are marketed for servicing agencies or in-house testing, but before an employer undertakes responsibility for their use in-house the operator should have the necessary training. Hand-held instruments such as these, for use with mains electricity, should themselves be designed to IEC 348 or 1010 safety standards.

Atmospheric conditions

Dust is a common byproduct of the handling, drying, processing and storage of agricultural produce. It can present a health hazard and a fire hazard. It can be abated in some circumstances by ionisation of the air, which can cause attraction and clustering of dust particles, by the induction process that is the basis of electrostatic spraying (section 4.3). The cluster then gravitates to the ground rather than remaining airborne. Automated equipment makes it possible to remove the operator from the scene, of course, provided that it is dust resistant to IP6 standard (section 2.4).

In poultry houses ammonia levels can rise above the HSE limit of 25 ppm, while CO_2 may exceed normal concentrations – although not to a hazardous extent. Portable chemical equipment can be used for spot checks on these levels. The application probably does not justify the use of an IRGA for CO_2 measurement but the development of a suitable ammonia sensor could be beneficial. Air temperature, ventilation and RH measurements in buildings are related to human welfare in some conditions. Machinery noise in buildings must also be considered. The output sound power, coupled with reflections within the buildings, can easily put the SPL over 85 dB(A) in many cases.

Malodour, like noise, is an atmospheric factor which can cause public reaction as well as adversely affecting workplace conditions. Reference was made in section 4.4 to the use of arrays of conductive polymer elements as odour sensors. These mimic the nose, with its array of chemical sensors, feeding information to the brain, which then learns to recognise the pattern (or signature) of particular odours. A 20-element polymer array, mounted in a 40-pin integrated circuit chip, with associated electronics, has been evaluated as a monitor for the odour from pig slurry. The elements in the array had different chemical selectivity and their responses (change of resistance) were mapped after signal processing. It was found possible to produce odour clusters analogous to the colour clusters described in section 4.3. The clusters can then be related to particular odour species. The objective of on-going research is to develop an automatic odour recognition system based on a trained artificial neural network.

7.4 Further information

In the UK more detail on the aspects of safety covered by this short chapter (and others that are not) can be obtained from the Health and Safety Executive via Stationery Office Publications (formerly HMSO) bookshops in London, Birmingham, Bristol, Manchester, Belfast and Edinburgh.

Textbooks on noise and vibration deal in depth with the many techniques and types of equipment used to measure and control these important quantities.

Publications on topics specific to this chapter include the following reviews and research papers.

Section 7.2

Vehicle stability

Hunter, A. (1992) A review of research into machine stability on slopes. *The Agricultural Engineer*, **47**, 49–53.

Owen, G., Hunter, A. & Glasbey, C. (1992) Predicting vehicle stability using portable measuring equipment. *The Agricultural Engineer*, **47**, 101–104.

Noise and vibration

Hilton, D.J. & Moran, P. (1975) Experiments in improving tractor operator ride by means of a cab suspension. *Journal of Agricultural Engineering Research*, **20**, 433–48.

ISO 2631 (1974) *Guide for the evaluation of human response to whole-body vibration.* International Organisation for Standardisation, Geneva.

ISO 5008 (1979) *Agricultural wheeled tractors and field machines – measurements of whole-body vibration at the operator*. International Organisation for Standardisation, Geneva.

Lines, J.A., Lee, S.R. & Stiles, M.A. (1994) Noise in the countryside. *Journal of Agricultural Engineering Research*, **57**, 251–61.

Suggs, C.W. & Huang, B.K. (1969) Tractor cab suspension design and scale model simulation. *Transactions of the American Society of Agricultural Engineers*, **12**, 283–9.

Controls

Sato, K., Hoki, M. & Solokhe, V.M. (1993) Voice recognition by neural network under tractor noise. *Transactions of the American Society of Agricultural Engineers*, **36**, 1223–7.

Section 7.3

Byun, H.G. Persaud, K.C., Khaffaf, S.M., Hobbs, P.J. & Misselbrook, T.H. (1997) Application of unsupervised clustering methods to the assessment of malodour in agriculture using an array of conductivity polymer odour sensors. In: Special issue: Livestock monitoring (ed A.R. Frost) *Computers and Electronics in Agriculture*, **17**, 139–261.

McLean, K.A. (1992) Grain dust-hazards, legislation, control. *The Agricultural Engineer*, **47**, 14–16.

Whyte, R.T. (1993) Aerial pollution and the health of poultry farmers. *World's Poultry Science Journal*, **49**, 139–56.

Appendix

A.1 The international system of units

These are divided into three groups.

Base units

- Length in metres (m)
- Mass in kilograms (kg)
- Time in seconds (s)
- Electric current in amperes (A)
- Thermodynamic temperature in kelvin (K)
- Amount of substance in mole (mol)
- Luminous intensity in candela (cd)

Supplementary units

- Plane angle in radians (rad)
- Solid angle in steradians (sr)

Derived units (selected)

- Frequency in hertz (Hz) /s
- Force in newtons (N) kg m/s^2
- Pressure in pascals (Pa) N/m^2
- Energy, work or quantity of heat in joules (J) N m
- Power or radiant flux in watts (W) J/s
- Electric charge in coulombs (C) A s
- Electric potential in volts (V) W/A
- Capacitance in farads (F) C/V
- Electric resistance in ohms (Ω) V/A
- Electric conductance in siemens (S) A/V
- Magnetic flux in weber (Wb) V s
- Magnetic flux density in tesla (T) Wb/m^2
- Inductance in henrys (H) Wb/A

- Celsius temperature in degrees celsius (°C) (°C = K – 273.15)
- Luminous flux in lumens (lm) cd sr
- Illuminance in lux (lx) lm/m^2
- Luminance in candela/m^2 (cd/m^2)

A.2 Other units

- bits – binary digits
- bauds – bit rate (bit/s)
- bytes – 8-bit groups (Kb = 1024 bytes, which is the closest binary equivalent to 1000)
- dB/dB(A) – decibels/decibels, A-weighted (section 7.2)
- eV – electron volts (Table 1.4)
- hectolitre weight (section 5.3)
- IP number – protection afforded by equipment enclosures (section 2.4)
- mcwb – moisture content, wet basis (section 5.3)
- pF – matric suction in soils (section 3.2)
- pH – acidity or alkalinity (section 1.16)
- ppm/vpm – parts or volumes/million
- RH/ERH(%) – relative humidity/equilibrium RH (sections 1.4 and 5.3)
- sps – steps/s
- wheelslip (%) (section 3.3)

A.3 Standardising and regulatory bodies

ANSI – American National Standards Institute
BIPM – International Bureau of Weights and Measures
BSI – British Standards Institute
CCITT – International Consultative Committee for Telegraphy and Telephony
CIE – International Commission for Illumination
DIN – Institute for Standardisation, Germany
DTI – Department of Trade and Industry, UK
EC – European Community (EN Standards and Directives)
EIA – Electronic Industries Association, USA
ETSI – European Telecommunications Standards Institute
HSE – Health and Safety Executive, UK
ICRPMA – International Committee for Recording the Productivity of Milk Animals
IEC – International Electrotechnical Commission
IEE – Institution of Electrical Engineers, UK
IEEE – Institute of Electrical and Electronic Engineers, USA

ISO – International Organisation for Standardisation
ITU – International Telegraphic Union
NBS – National Bureau of Standards, USA
NPL – National Physical Laboratory, UK
OIML – International Organisation of Legal Metrology
RTCM – Radiotechnical Commission for Maritime Services
SAE – Society of Automotive Engineers, USA

A.4 Glossary of terms

Accuracy The difference between the value of a quantity measured by an instrument and that determined by a standard method under the same conditions.

ACNV Automatic control of natural ventilation.

ACR Automatic cluster removal in milking parlours.

ADAS Agricultural Development and Advisory Service, England and Wales.

ADC Analogue to digital converter. A device for converting the level of an analogue signal to an equivalent digital representation at short, repeated intervals.

AI Artificial intelligence. The ability of a device to draw inferences from data and to respond on the basis of rules and experience.

Algorithm The mathematical or logical operation by which the output of a controller is derived from its input.

Amplitude ratio The ratio of the sensitivity of an instrument or instrument component at a particular signal frequency to its d.c. sensitivity.

Analogue A representation of a variable by a physical quantity (e.g. a voltage) which is made proportional to the variable.

ANN Artificial neural network. A device that can be trained to recognise specific features in input data and to classify the data on that basis.

ASCII American standard code for information exchange (equivalent to ISO 646).

ASIC Application specific integrated circuit.

Bus A group of wires, common to various units in a digital system, and used to carry data between them.

CA Controlled atmosphere in sealed crop stores.

CAN Controller area network. A bus system devised for vehicle applications.

CCD Charged coupled device. A form of digital camera.

Closed loop control Control action initiated by feedback from the controlled process.

CMRR Common mode rejection ratio. The ability of an electronic module or system with a two-wire input to prevent a loss of accuracy in the presence of unwanted signals induced equally in both conductors.

CPU Central processor unit of a computer, which performs the arithmetic and logical functions, via access to the program being executed and the system's memory devices.

CRC Cyclic redundancy check. One form of diagnosis for errors in data transmissions.

Critical damping The degree of viscous damping of a resonant instrument, or instrument component, which is just sufficient to prevent an oscillatory response to changes at the input.

CRT The common display unit in TV equipment and VDUs.

CW Continuous wave (of electromagnetic radiation at radio and microwave frequencies).

DAC Digital to analogue converter. A device which produces an analogue presentation of data when supplied with data in digital form.

Damping ratio The ratio of the damping of an instrument or instrument component to its critical damping.

DDE Dynamic data exchange. A feature of some computer systems which allows the exchange of data between software applications.

DGPS Differential GPS. A technique for improving the accuracy with which a mobile GPS receiver can establish its position, through corrections supplied by a local base station.

Dielectric An electrical insulator. In an electrical field its electrons and ions become separated and the material is polarised in the direction of the field. This results in surface charges of opposite sign on opposing surfaces perpendicular to the field. Polarisation can also be caused in some dielectrics by mechanical stress.

Dielectric constant When the space between the plates of a capacitor is filled with a dielectric material its capacitance is increased by a factor commonly known as the dielectric constant of the material. This factor is identical to the property of the material known as its relative permeability.

DSP Digital signal processor. A computer dedicated to intensive mathematical calculations.

EC Electrical conductivity in siemens.

ECU Electronic control unit.

Electrolyte A salt or acid solution in which current is carried by ions.

EMC Electromagnetic compatibility. The ability of electronic equipment to operate in an electrically noisy environment and to do so without creating EMI.

EMI Electromagnetic interference. Radiated electrical noise.

EPROM Erasable programmable read-only memory. A device with memory that can be erased by ultraviolet light or by applying a suitable electrical signal.

Error In a control system, the difference between the set point and a measured value of the controlled quantity.

Ethernet One form of local area network.

Feedback In a control system, the transmission of information from the output to the input of the system or a section of it.

Feed forward In a control system, a means of signalling a change at an input to later changes of the system.

FET Field-effect transistor. A special form of the transitor with a high input impedance and low current requirement.

FM Frequency modulation. An electrical system in which the frequency of an alternating (carrier) signal varies in proportion to the value of the variable that it represents.

Frequency response The manner in which the sensitivity of an instrument, or instrument component, varies with a.c. inputs of different frequency.

fro Full-range output. The output of an instrument or instrument component at the upper end of its working range.

Fuzzy logic A means of analysis and control where variables are not quantifiable numerically but are replaced by linguistic variables such as *low*, *medium* and *high*.

GA Genetic algorithms. A search method for optimal control settings (*inter alia*) which employs analogues of the processes of natural selection.

GIS Geographical information system. A compilation of data on soils, crops, etc. related to their geographical position.

GPIB General purpose interface bus (aka IEEE488). A widely used standard for instrumentation systems.

GPR Ground-penetrating radar for subsurface location of discontinuities.

GPS Global positioning system. The USA's NAVSTAR satellite constellation for global navigation and position fixing.

Grey scale In digital image analysis, the stepped scale in which the brightness of each picture element (pixel) is classified.

GUI Graphical user interface. VDU displays for interaction with the computer user.

Handshaking A method of controlling data flow when transmitters and receivers do not operate at the same speed. The data flow is controlled by the receiver.

Hysteresis The amount by which the output from an instrument, or instrument component, can differ at any point in its working range, depending on the direction of change at its input (i.e. increasing or decreasing). Hysteresis is usually referred to its magnitude at $\frac{1}{2}$ fro, expressed as a percentage of fro.

IGBT Isolated gate bipolar transistor. Power switching transistor.

Impedance Total effective resistance of an electric circuit to alternating current. A combination of ohmic resistance, capacitance and inductance. (Inverse quantity, admittance.)

Interface The common boundary between units of a system, across which data can be transferred. The interface is required to make units compatible with respect to codes, operating speeds, signal levels, etc.

Ions Atoms which have lost or gained electrons, leaving them positively or negatively charged, respectively, due to the positive charge on the nucleus.

IRGA Infra-red gas analyser.

ISA Internal systems adapter. Computer bus.

ISDN Integrated services digital network. High-speed, multimedia communications link.

ISE Ion-selective electrode for ionic measurements in liquids.

ISFET Ion-selective FET. Semiconductor equivalent of the ISE.

LAN Local area network.

LCD Liquid crystal display unit.

LED Light-emitting diode.

Linearity The degree to which the sensitivity of an instrument, or instrument component, remains the same over its working range. This is usually expressed as the maximum departure from linearity anywhere in that range, as a percentage of fro.

LSB Least significant bit in a binary number.

LVDT Linear variable differential transformer. Sensor for measurement of position and displacement.

MAC Manufacturing automation control network introduced by General Motors.

MMI The man–machine interface.

Modem Device which encodes serial digital data for transmission as an audiofrequency signal over the telephone network and decodes incoming data back into serial digital form.

MRI Magnetic resonance imaging for non-destructive internal inspection of hydrogen-rich material in particular.

MSB Most significant bit in a binary number.

MUX Multiplexer. Device for switching signals, in sequence, over a common set of wires.

NIR The near infra-red band of the electromagnetic spectrum.

Noise (electrical) Undesirable signals (interference) which distort or mask desirable signals.

Non-volatile memory Memory system which does not lose its data contents if the power is removed (e.g. ROM).

NTC Negative temperature coefficient (of a resistance element). Characteristic of some semiconductor temperature sensors, such as the thermistor. Electrical resistance decreases as their temperature increases, unlike metal resistance.

OLE Object linking and embedding. Microsoft software which treats files, spreadsheets, etc. as *objects*. Intended to make software less vendor-dependent (i.e. more *open*).
OPC OLE for process control.
OSI ISO's Open Systems Interconnection model for *open* networks.

PAR Photosynthetically active radiation in the electromagnetic spectrum (400 to 700 nm).
Parity check A method of error checking in which an extra bit is added to a data frame to make the total number of 1s odd or even, according to a chosen convention.
PAT Portable appliance tester for electrical safety checks.
PC Personal computer (desk-top machine).
PCI Peripheral component interface bus.
PCMCIA Personal computer memory card International Association interface cards with a variety of functions.
Peripheral A device external to the processor, such as a printer, floppy disk unit or VDU, which is controlled by and communicates with the CPU.
PF Power factor in an a.c. circuit. PF = 1 when the sinusoidal voltage and current in the circuit are in phase. Capacitative and inductive elements in the circuit produce an out-of-phase angle, ϕ, and the PF is reduced by a factor cos ϕ, representing energy dissipation.
Photon A *particle* of electromagnetic radiation.
PID The proportional + integral + differential control algorithm.
Pixel In digital image analysis, a single picture element, with a classified brightness and/or colour value.
PLC Programmable logic controller.
Port Communications inlets and outlets such as the COM 1, COM 2, etc. connections to a PC, with settings for specific purposes (modems, etc.).
Precision (or discrimination) The smallest change in a measured quantity that produces a perceptible change in the output of an instrument. It is not to be equated to accuracy.
Protocols These establish the procedure through which data communication takes place. Senders and receivers must employ the same protocol.
PRT Platinum resistance thermometer, used as a working standard for temperature measurement.
PWM Pulse width modulation of an alternating signal. One means of communicating digital data.

RAM Random access memory. Read/write memory which provides the necessary data storage during computer processing.

Range The working range of an instrument, over which its response to changes in a measured quantity follow a given law.

RCD Residual current safety cut-out in an electrical system.

Real time A system operates in real time when data generated by an event are processed virtually simultaneously.

Repeatability The variation in the results obtained when a measurement is repeated, under the same conditions, as far as possible.

RF The radiofrequency band of the electromagnetic spectrum.

RFI Radiofrequency interference.

RFID Radiofrequency identification system (e.g. for livestock identification).

RGB The red, green and blue colour bands used in colour television, etc.

RLL Relay ladder logic for programming of PLCs.

rms Root mean square value of an alternating signal.

ROM Read only memory. A memory that has a permanent data pattern written into it.

SCADA Supervisory control and data analysis multilevel system.

SCR Silicon-controlled rectifier. A unidirectional current switch controlled by a *gate* electrode.

Segmentation In image analysis, the process of identifying areas in the image with specific features, such as colour values.

Semiconductor A material whose electrical conductivity is intermediate between that of metals and that of insulators.

Sensitivity The change in output from an instrument or instrument component, per unit change at its input.

Set point In a control system, the desired value of the controlled quantity.

SLM Sound level meter for noise measurement.

SOM Soil organic matter.

Span Range of an instrument display.

SPL Sound pressure level, in dB.

Step response The response of a resonant instrument, or instrument component, to a step change at its input.

SVAPS Spatially variable agricultural production systems. One name for precision farming techniques.

TCP/IP Transport control protocol/Internet protocol. Protocols which allow computers to share resources across the Internet.

TDR Time domain reflectometry. A method of detecting dielectric discontinuities, employed in soil moisture measurement.

Thresholding In image analysis, the removal of detail outside the segmented area(s) by setting all the background pixels to the same brightness (usually either black or white).

Thyristor Another name for the SCR.

Time constant A quantity characterising the response of a non-resonant instrument or instrument component to a step change at its input.

Traceability The hierarchical calibration process through which the readings of an instrument are referred to the appropriate international standard.

Transducer (electrical) A device which produces an electrical output in response to a specific physical quantity, property or condition which is measured.

Transient response The response of a resonant instrument, or instrument component, to a short impulse at its input.

Transponder A device that yields information when interrogated by an external instrument, such as the passive (battery-less) units used for automatic animal identification.

Triac The equivalent of two SCRs connected in inverse parallel. Used as a bidirectional current switch, as in soft-start units for electric motors.

UHF The ultra-high-frequency region of the electromagnetic radiation spectrum.

UPS Uninterruptible power supply, for protection of electrical and electronic equipment from mains-borne interference and to bridge periods when the mains supply is off.

VDU Visual display unit in computer systems and elsewhere.

Verification Confirmation of stated accuracy of an instrument reading or a measurement.

Volatile memory Memory, such as conventional RAM, which loses its contents if power is removed.

WAN Wide area network.

WGS World Geodetic System. Model of the earth's shape which is employed to convert GPS data to latitude, longitude and altitude.

WWW The WorldWideWeb. A major source of information on the Internet, which can be accessed via service providers and explored with the aid of browsers and search engines.

Index